Molecular Phylogenetics

K. Praveen Karanth

Molecular Phylogenetics

Building Evolutionary Trees using DNA Sequence Data

Springer

MANIPAL
UNIVERSAL PRESS

K. Praveen Karanth
Centre for Ecological Sciences
Indian Institute of Science
Bengaluru, Karnataka, India

ISBN 978-981-95-9146-6 ISBN 978-981-95-9147-3 (eBook)
https://doi.org/10.1007/978-981-95-9147-3

Jointly published with Manipal Universal Press

The edition is not for sale in India. Customers from India should order the print book from Manipal Universal Press (MUP). ISBN of the India edition: [978-93-47805-00-4].

This Springer imprint is published by the registered company Springer Nature Singapore Pte Ltd.
The registered company address is: 152 Beach Road, #21-01/04 Gateway East, Singapore 189721, Singapore

Preface

In the last two decades, there has been an astronomical increase in DNA sequence data in various public databases. In recent times, complete genomes of numerous organisms have also been sequenced. Additionally, there have been tremendous advances in sequence analyses methods and computational power of computers. These developments have contributed towards the explosive growth in phylogenetics studies. Today, labs across the world routinely employ molecular phylogenetic methods in their research. I have been using molecular phylogenetic approaches in my research for over a decade and a half. For me, it has been an exciting learning period but also a very frustrating one due to the lack of a good textbook on molecular phylogenetics. Numerous books that cover various aspects of molecular phylogenetics are available, but one has to glean relevant information from different books to get the complete picture. The inspiration to write this book stems out of the frustration that all relevant information regarding phylogenetics is not available in a single book. Additionally, I felt that there was a need for a textbook that covers the fundamentals of molecular phylogenetics, without going into excruciating mathematical details. A textbook that is conceptually accessible to taxonomists, biogeographers, evolutionary biologist and biologists in general. A book that would help one understand, in simple terms, what the phylogeny programs are really doing, i.e., the theoretical underpinnings of the phylogenetic approach and various tree-building methods. I hope this book will achieve this goal of providing a brief and simple overview of phylogenetics. Furthermore, I hope that this book will help students of phylogenetics analyse and interpret data correctly and understand some of the issues pertaining to phylogenetics analysis. However, I must point out that I have oversimplified the description of some of the methods. This has been done deliberately to keep things simple. For more detailed and accurate descriptions of the various methods, I would urge the readers to refer to the original sources provided in this book.

This book does not delve into the molecular tools used to generate sequences or programs used to analyse sequence data. It is work in progress and periodically I hope to bring out updated issues that will make the content as current as possible. The book is particularly targeted towards Indian researchers doing molecular systematics, biogeography, phylogeography, character evolution and evolution in general. This is the first draft of a work in progress, more like class notes than a book. The references are not up to date and in some cases I am sure more appropriate papers could have been cited. Most of the examples I have discussed come from

Indian systems and the work done in my lab. I hope to update all this in future editions. I would like to take this opportunity to thank Mihir, and Chinta for taking the time to proofread the material. Without their help, the book would not have attained its current form. Thanks to Hemanya and Chinta for corrections on the previous version and Shabuddin Shaik for help with figures. Big thanks to Prof. R. Geeta for reviewing the book, her thoughtful comments and suggestions have significantly improved the book. Comments, suggestions and corrections are welcome.

Bengaluru, Karnataka, India K. Praveen Karanth

Competing Interests The author has no competing interests to declare that are relevant to the content of this manuscript.

About the Book

Molecular Phylogenetics: Building Evolutionary Trees Using DNA Sequence Data has expanded dramatically over the past four decades. Today, researchers across almost every branch of biology generate DNA sequence data and perform at least some basic phylogenetic analyses. Yet much of this work relies on 'user-friendly' software with pull-down menus, leaving users with a limited understanding of the methods being applied behind the screen.

This book fills that gap by offering a clear conceptual framework for the process of tree building, enabling practitioners to make more informed and effective use of the analytical tools available to them. It begins with a historical overview of tree construction and explores the deep connections between taxonomy, systematics and phylogenetics. Various tree-building methods are presented with straightforward explanations of their advantages and limitations, supported by explicit examples.

Key aspects of phylogenetic inference and interpretation are also examined, including tree rooting, branch-length estimation, bootstrap support, hypothesis testing, model selection and more. Important challenges, such as discordance between molecular and morphological data, as well as conflicts among gene trees, are addressed. The penultimate chapter highlights diverse applications of phylogenetics, followed by a helpful chapter on frequently asked questions.

Written in a simple and accessible style, this book allows non-specialists to grasp essential concepts with ease. It will be valuable to anyone interested in evolution, with or without formal training in biology.

Contents

About the Author

K. Praveen Karanth did his BSc in Agriculture at University of Agricultural Sciences, Bangalore. After which he went to do his PhD from State University of New York Albany. His PhD thesis dealt with phylogeny and biogeography of Asian leaf monkeys. After four years of post doc at Haifa University, Israel, and Yale University, USA, he returned to India to take up a position in IISc, Bangalore. He is currently a Professor at the Centre for Ecological Sciences, IISc. His research area deals with the application of phylogenetic and population genetic tools in understanding micro- and macro-evolutionary processes. He is particularly interested in biogeography, character evolution, phylogenetic comparative methods, molecular systematics and conservation genetics.

Deconstructing a Phylogeny

1

Abstract

This chapter introduces the fundamental structure of phylogenetic trees, explaining various terms used to describe them, such as nodes, branches, roots, clades and sister taxa. It contrasts cladograms and phylograms, clarifying how branch lengths either lack meaning (cladogram) or represent evolutionary distance (phylogram). The concepts of rooted and unrooted trees are also introduced. Two methods of rooting trees, namely, midpoint and outgroup rooting, are discussed. The terms 'monophyly', 'paraphyly' and 'polyphyly' are defined, and their usage in describing the incongruence between traditional taxonomy and molecular phylogeny is explored. This foundation sets the stage for subsequent chapters.

Keywords

Cladogram · Phylogram · Unrooted tree · Clade · Paraphyly · Polyphyly

Phylogeny is an evolutionary tree that shows how species are related to each other. For example, Fig. 1.1 is a phylogeny showing relationships between five species. A phylogeny consists of branches and nodes. Internal nodes (6–8) represent the common ancestors that have undergone speciation and external nodes represent the extant (living) species. The common ancestor of all the five species is represented by the root node (9). Branches are lines that connect nodes. In this phylogeny, the five species fall into two clusters or clades; species 1, 2 and 3 are members of clade A, and 4 and 5 belong to clade B. A clade constitutes the common ancestor and all its descendants. Members of each clade are more closely related to each other than to the members of other clades. For example, species 1 and 2 are closely related (sister taxa), i.e. share an immediate common ancestor (6), whereas species 1 and 5 are distantly related and share a ('distant') common ancestor at the root node (9).

There are two ways to represent a phylogeny: a cladogram and a phylogram. A cladogram shows how species are related to each other; branch lengths do not have any meaning here as they do not correspond to genetic distance (Fig. 1.2a). Thus, a cladogram only tells us how species are related and which nodes represent the

K. P. Karanth, *Molecular Phylogenetics*, https://doi.org/10.1007/978-981-95-9147-3_1

Fig. 1.1 A phylogeny showing internal and external nodes

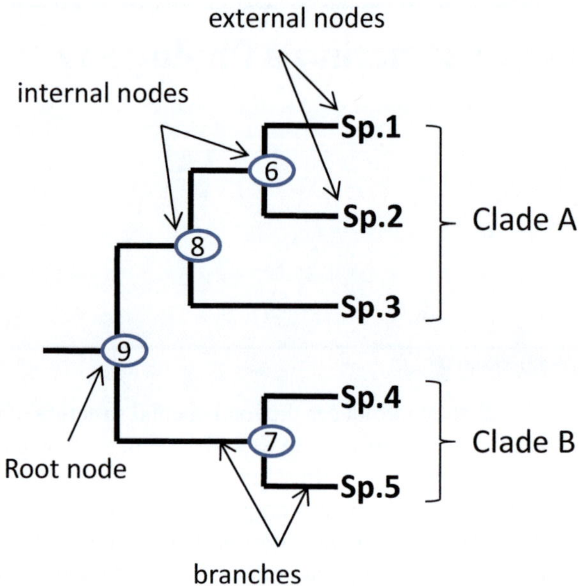

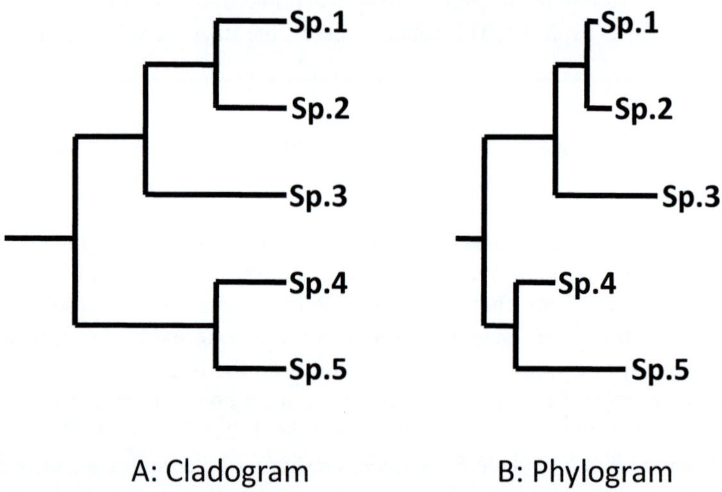

Fig. 1.2 A phylogeny showing internal and external nodes

immediate common ancestors of the extant species, but does not provide the genetic distances between these species. On the other hand, in a phylogram (Fig. 1.1b), branch lengths correspond to the number of changes along a branch (lineage). Here, one can actually get a visual representation of the genetic distance between species. For example, we know that species 1 and 2 are sister species, and so are species 4 and 5. However, the evolutionary distance (genetic distance) between 1 and 2 is

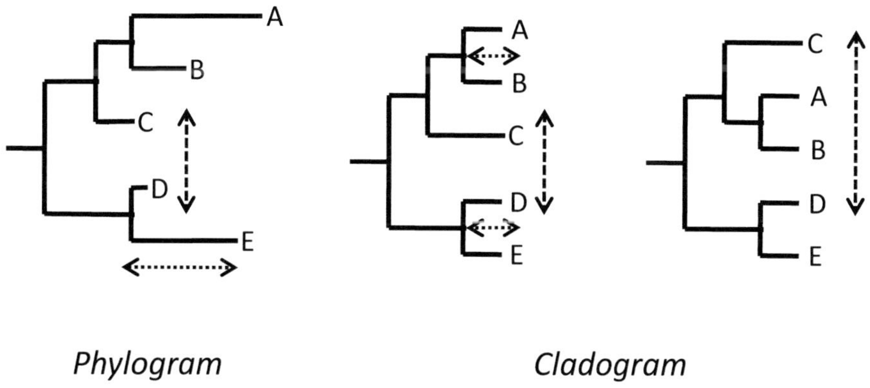

Fig. 1.3 Contrast between a phylogram and a cladogram. Dotted and dashed lines represent horizontal and vertical distances, respectively

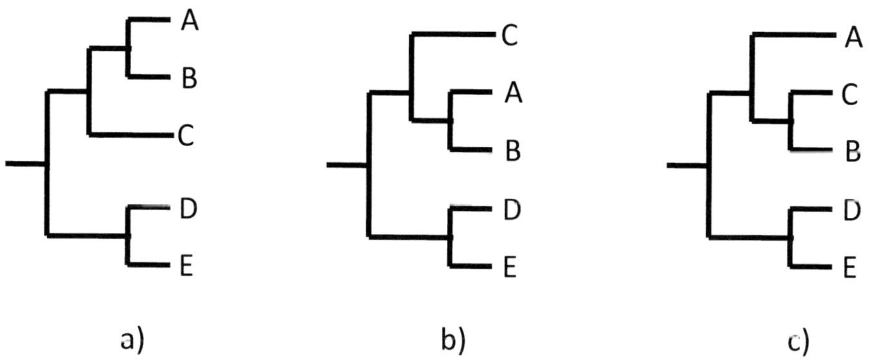

Fig. 1.4 Cladograms showing similar (**a** and **b**) and different (a/b vs. c) tree topologies

much less than that between species 4 and 5. This information is not provided by a cladogram.

In both a phylogram and a cladogram, vertical distances (dashed line in Fig. 1.3) have no meaning, i.e. they do not provide any information. For example, the vertical distance between species C and D is not the same in the two cladograms, but this does not mean that these two taxa are more closely related to each other in one cladogram than in the other.

Thus, vertical distances between taxa have no meaning in both a cladogram and a phylogram, and horizontal distance has no meaning in a cladogram.

Phylogeneticists often use the term 'tree topology', which reflects how species are related to each other in an evolutionary tree. For example, the two cladograms in Fig. 1.4a, b have the same topology even though visually the arrangement of the taxa A, B and C appears different. Whereas the tree topology of cladograms b and c is different even though visually the branching patterns of the two trees appear similar. In Fig. 1.4b, species A and B share an immediate common ancestor, whereas in Fig. 1.4c, species C and B share an immediate common ancestor.

Rooted and Unrooted Trees

Till now we have been dealing with rooted trees. A rooted tree is one where the common ancestor of all extant species is shown as the root of the tree. Of course, this common ancestor has long gone extinct. In Fig. 1.1, the root of the tree is represented by node 9, which is the common ancestor of extant species A–E. Rooted trees can either be depicted as a rectangular cladogram (Fig. 1.5a) or a slant cladogram (Fig. 1.5b).

However, all tree-building methods, except unweighted pair group method with arithmetic mean (UPGMA), generate unrooted trees (*see* Chap. 2 for UPGMA). In Fig. 1.5, both rooted and unrooted trees are shown, and these trees have the same topology. However, in the case of an unrooted tree (Fig. 1.5c), the point (node) that represents the common ancestor of the four extant species is not known. If we have four species, then one can come up with three possible unrooted trees depicting their relationships (Fig. 1.6).

Each unrooted tree has five branches and two internal nodes. Depending upon where (which branch) the root is placed, we get different rooted trees. For example, if we take the unrooted tree depicted in Fig. 1.6a, the root can potentially be placed on any one of the five branches, yielding five different topologies.

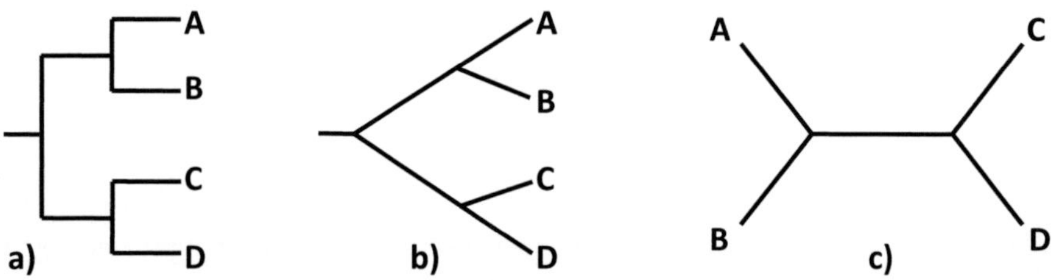

Fig. 1.5 Rooted trees (**a** and **b**) and an unrooted tree (**c**)

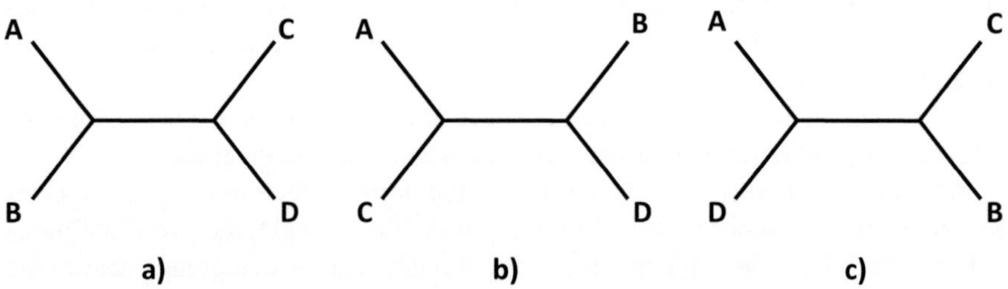

Fig. 1.6 Three possible unrooted trees showing relationships between four taxa

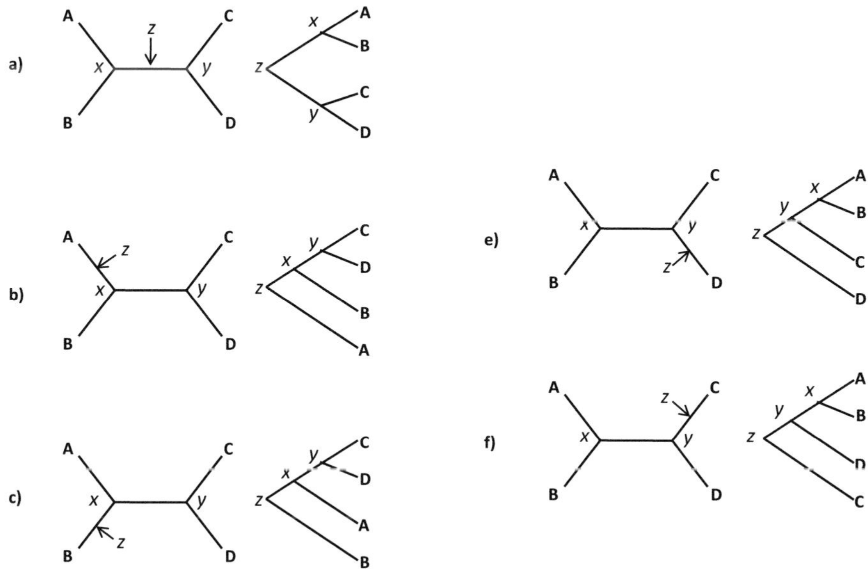

Fig. 1.7 Five possible rooted trees for a given unrooted tree (Fig. 1.6a). Arrows indicate the location of the root. Internal nodes are denoted by x and y, and the root node by z

Figure 1.7 shows the five possible rooted trees for a given unrooted tree (from Fig. 1.6a). In each of these cases, the root (arrow) was placed in a particular branch, thereby giving the corresponding rooted tree. For example, in the case of Fig. 1.7a, the root was placed along the branch connecting internal nodes x and y. Given that in the case of four-taxa phylogeny there are 3 possible unrooted trees and each unrooted tree can be rooted in 5 different ways, we get a total of 15 possible rooted trees.

Rooting Unrooted Trees

To convert an unrooted tree to a rooted tree, we bisect the unrooted tree at the point where the root is placed (Fig. 1.8). This gives us two clades, one to the left of the root and the other to its right. The branches leading to the common ancestors of these clades are then joined to give the root such that all the extant taxa line up to the right of the root (Fig. 1.8).

A rooted tree is more informative than an unrooted tree. Firstly, in a rooted tree, the direction of evolution is clearly established; it is from the root towards the tips of the tree. Secondly, the relationships between species are better resolved in a rooted tree. For example, take the unrooted tree depicted in Fig. 1.8b. This unrooted tree suggests that species A and B form one cluster and species C and D form another. Thus, the unrooted tree suggests that species A and B are sister taxa.

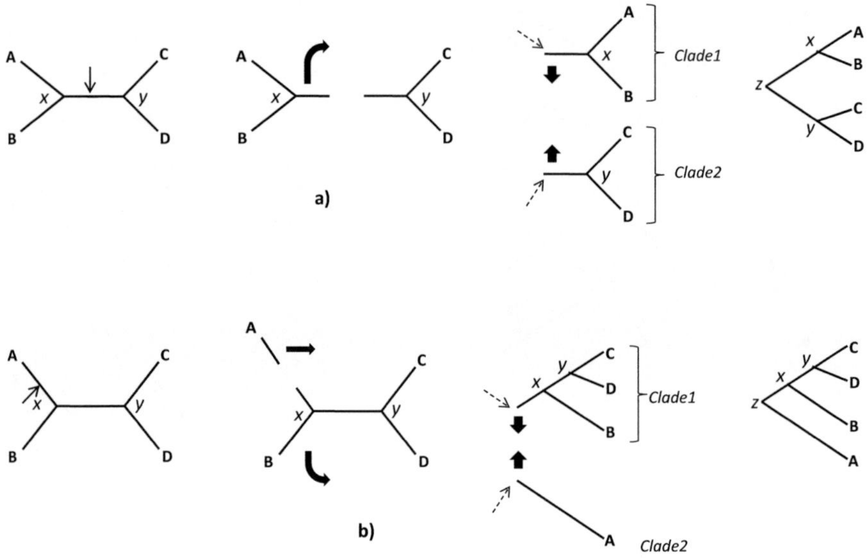

Fig. 1.8 Rooting an unrooted tree. Dashed arrows indicate the branches leading to the common ancestors of these clades 1 and 2

However, on rooting this tree, we realise that species B is more closely related to C and D than to species A. Thus, we have better clarity regarding relationships between species in the case of rooted trees.

As mentioned earlier, most tree-building methods generate unrooted trees. The researcher has to then root the trees using an appropriate method. Determining the correct location of the root is an important component of phylogenetic analysis. This is apparent from the example discussed above, where for a given four-taxa unrooted tree, one can generate five different rooted trees. Each of these trees produces a different phylogenetic hypothesis. There are two ways to determine the root of a tree: outgroup rooting and midpoint rooting.

Outgroup Rooting

In a phylogenetic study, we are interested in determining the relationships among certain taxa. These taxa could be species belonging to a particular genus, genera belonging to a family and so on. This taxonomic group of interest is called the 'ingroup'. We are basically interested in the relationships among the members of the ingroup. For example, we might be interested in the phylogenetic position of humans among the apes. In this case, human along with apes (chimp, gorilla, orang and gibbon) is our ingroup. To root this tree, we also include an 'outgroup'. The outgroup constitutes a taxonomic group that is related to the ingroup but is not part of the ingroup. Ideally, it is sister to the ingroup. From traditional taxonomy, we know that the old-world monkeys (OWM) are a sister group to the apes + human

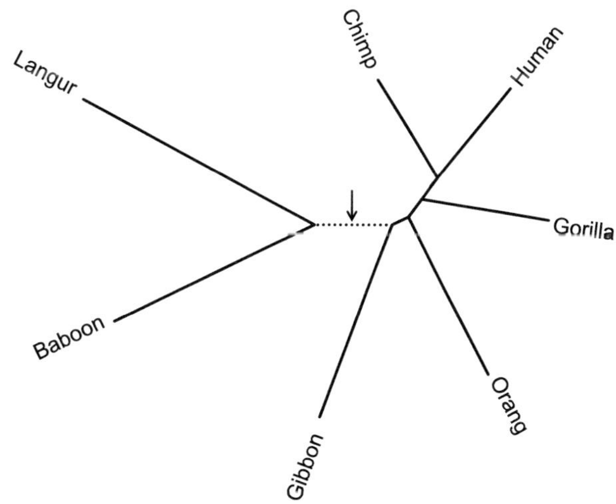

Fig. 1.9 Unrooted tree showing relationships between humans and apes. The dotted line represents the branch connecting the outgroup (langur and baboon) and the ingroup. The arrow indicates the location of the root

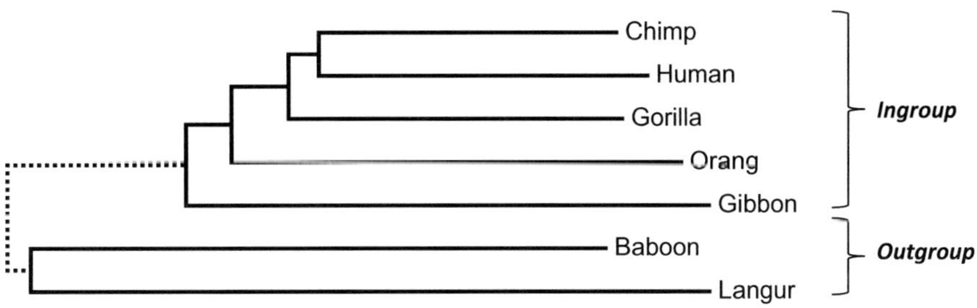

Fig. 1.10 Rooted tree showing relationships between humans and apes. The dotted line represents the branch connecting the outgroup and the ingroup

group. Thus, we can use OWM, such as baboons and langurs, as an outgroup in our analysis. It is always a good idea to include multiple outgroups.

The example discussed above is illustrated in Fig. 1.9, which is an unrooted tree showing relationships among members of the ingroup. Here, langur and baboon represent the outgroup. The root of this tree, depicted by the arrow, is placed on the midpoint of the branch (dotted line) that connects the ingroup with the outgroup. The unrooted tree is then redrawn using the method illustrated in Fig. 1.8 to give us the rooted tree (Fig. 1.10). It must be noted that the outgroup should not be very distantly related, as it could cause distortion in the topology of the ingroup. In our example, a very distantly related outgroup would be a non-primate mammal, such as a rat. Similarly, an outgroup that is very closely related to the ingroup can also generate erroneous topology because it might be a member of the ingroup.

In our example, the outgroup could be chosen without much difficulty because primate taxonomy is better understood than most other groups. In taxonomically problematic groups, it is always a good idea to include different related taxa as potential outgroups to enable outgroup rooting. However, in some cases finding a suitable outgroup might be difficult; in such cases, one can use midpoint rooting.

Midpoint Rooting

In the absence of a suitable outgroup, we first generate an unrooted phylogram of the ingroup. Next, the branches representing the longest tree distance between any two taxa are identified. The root is placed at the midpoint of the sum of these branches. In our unrooted tree, the longest tree distance is between orang and gibbon (Fig. 1.11). The arrow represents the midpoint between these two taxa. Once the root is placed, the unrooted tree is redrawn using the method illustrated in Fig. 1.8 to give us the rooted tree (Fig. 1.12). Midpoint rooting assumes molecular clock, i.e. the rate of molecular evolution along various branches is equal. This assumption is almost always violated. Therefore, midpoint rooting should be used as a last resort as this method can generate erroneous topology when rates of molecular evolution vary across the phylogeny.

It must be noted that most phylogeny programmes will display a rooted tree on the computer screen even though the tree building algorithm has generated an unrooted tree. We do not get to see the unrooted tree unless one specifically asks for it. Most programmes, by default, root the tree by, usually, picking the first or the last

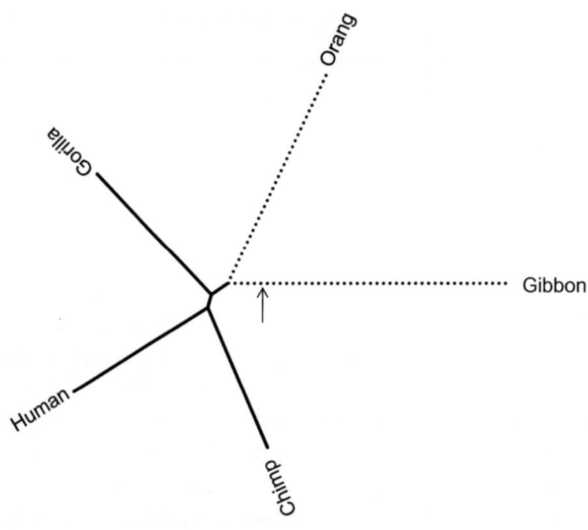

Fig. 1.11 Unrooted tree showing the relationships between humans and apes. The dotted line represents the longest tree distance between two taxa. The arrow indicates the location of the root

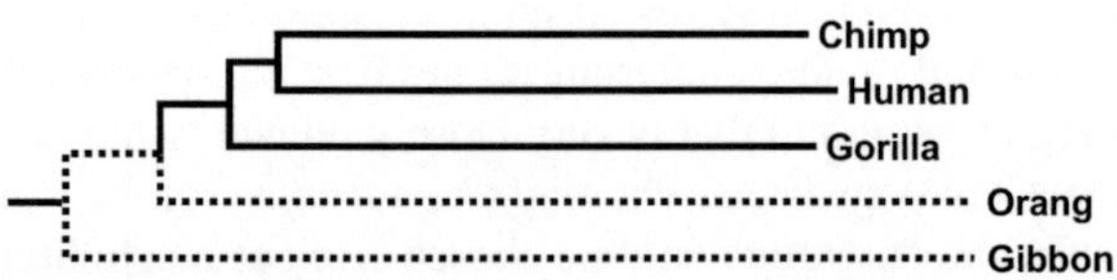

Fig. 1.12 Rooted tree showing the relationships between humans and apes: The dotted line represents the longest tree distance between two taxa

taxon in the alignment as outgroup and display a rooted tree. Alternately, the algorithm might have used midpoint rooting. If the taxon chosen by the programme is an incorrect outgroup, then the resulting topology will be erroneous. *Thus, do not accept the displayed tree as the final tree. Always* check to see whether the appropriate outgroup has been used. The outgroup should be sister to all other species in the phylogeny (ingroup). If the outgroup is not in the right place, then re-root the tree with an appropriate outgroup. Alternately, create an alignment wherein the outgroup taxon is either the first or the last sequence, depending on the programme. When undertaking a phylogenetic study, include potential outgroups in your sampling strategy if outgroup sequences are not available in GENBANK.

Describing Incongruence Between Traditional Taxonomy and Phylogeny

In phylogenetic studies, we often come across the terms 'monophyly', 'paraphyly' and 'polyphyly'. These are terms used to describe the status of taxa when the taxonomy of those taxa is compared with their molecular phylogeny.

Take the example shown in Fig. 1.13; here we are discussing the status of genus A, which consists of multiple species (A1, A2 and A3). When we build the molecular phylogeny of this species group (genus), three different outcomes are possible. Firstly, species in this genus might all branch together as expected from the taxonomy (Fig. 1.13a). Here, members of genus A constitute a natural group or a clade and we call genus A monophyletic. Similarly, genus B is also monophyletic and forms a sister clade. A clade is a monophyletic group that consists of all the descendants of an ancestral node (species). Members of a clade, by definition, form a cluster (are monophyletic); thus, the usage 'monophyletic clade' is incorrect. The phylogeny depicted in Fig. 1.13a is consistent with the taxonomy. In the second scenario, genus B is monophyletic; however, it is nested within a larger group consisting of members of genus A (Fig. 1.13b). In this phylogeny, genus A is

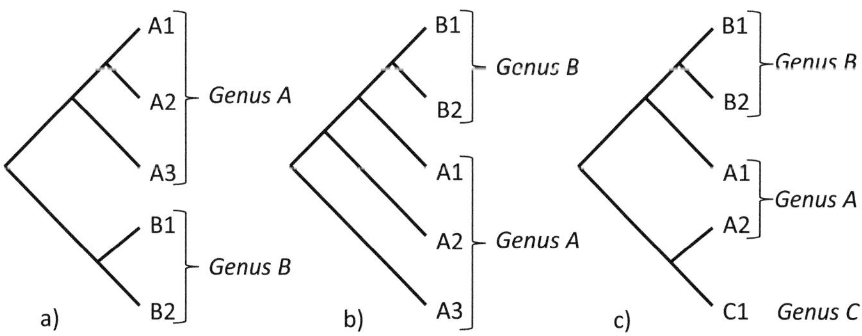

Fig. 1.13 Taxonomic status of a genus with respect to phylogeny. (**a**) Genus A is monophyletic. (**b**) Genus A is paraphyletic. (**c**) Genus A is polyphyletic

paraphyletic as it does not include species from genus B. In the third scenario, members of genus A are branching with different genera (B and C). We call this condition polyphyly, i.e. genus A is polyphyletic. In both the second and the third scenarios, genus A does not constitute a natural group or form a clade, and the taxonomy is not congruent with the phylogeny.

Under what conditions can we expect such incongruence? Paraphyly and polyphyly are often artefacts of incorrect taxonomy, i.e. unrelated species have been placed in the same group (in the same genus in our case). Taxonomy is often based on overall similarity of external morphological characters. These similarities might be due to homoplasy and not homology and therefore do not reflect the evolutionary relationships (see Chap. 2 for details). Take the example of Old and New World vultures. Superficially, they look very similar; however, the molecular phylogeny (Fig. 1.14a) suggests that they are not sister taxa and therefore do not form a natural group (clade). In this case, what we call 'vultures' is a polyphyletic group. The similarity between Old and New World vultures is due to the convergence of morphological characters.

Similarly, the molecular phylogeny of apes and humans does not support the monophyly of apes. Thus, apes do not constitute a natural group; on the other hand, they are paraphyletic as traditionally the taxa 'ape' did not include human (Fig. 1.14b). The overall similarity among apes is due to the retention of symplesiomorphies among them, the loss of these characters on the lineage leading to humans and the acquisition of autapomorphies along the human lineage (*see* Chap. 2). Nevertheless, it must be pointed out that today humans are considered apes.

What does one do when the phylogeny retrieves para/polyphyletic taxa? One obvious step is to revise the taxonomy, such that it is consistent with the phylogeny. In the example illustrated in Fig. 1.13b, genus B could be subsumed into A. Similarly, in the case of 13c, species A1 and A2 can be synonymised into genera B and C, respectively. This should be followed by a thorough revaluation of morphological characters used in the taxonomy of these genera. Nevertheless, it must be noted that in some cases, the phylogeny itself might be incorrect, thereby creating this

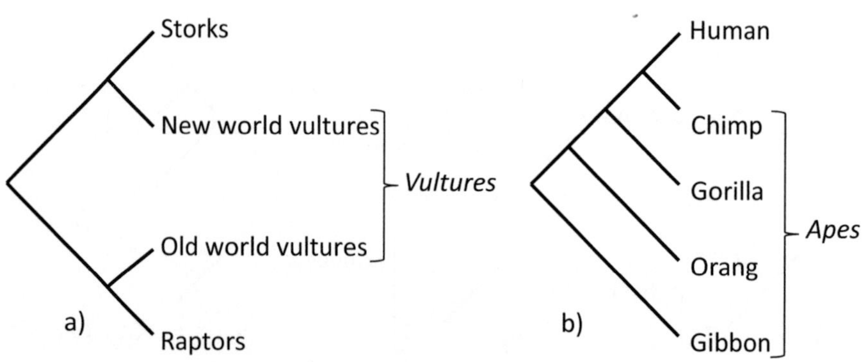

Fig. 1.14 Taxonomic status of apes and vultures. (**a**) Vultures are polyphyletic. (**b**) Apes are paraphyletic

incongruence between taxonomy and phylogeny. For example, the marker used for building the phylogeny might be under positive selection and therefore might not give the right phylogeny (*see* Chap. 8). Thus, it is always a good idea to derive multiple phylogenies from different markers as opposed to using a single phylogeny based on a single marker. For a more detailed treatment of some the topics discussed in this chapter see Page and Holmes (1998).

References

Page RDM, Holmes EC (1998) Molecular evolution: a phylogenetic approach. Blackwell Science, Oxford

Some Reference Material Dealing with Molecular Phylogenetics

Avise JC (1994) Molecular markers, natural history, and evolution. Chapman & Hall, New York
Avise JC (2000) Phylogeography. Harvard University Press
Felsenstein J (2003) Inferring phylogenies. Sinauer Associates, Sunderland
Graur D, Li W-H (2000) Fundamentals of molecular evolution, 2nd edn. Sinauer Associates Inc., Sunderland
Hall BG (2018) Phylogenetic trees made easy: a how-to manual (V edition, with CD-Rom). Oxford University Press
Hillis DM, Moritz C, Mable BK (1996) Molecular systematics. Oxford University Press, p 1058
Lemey P, Salemi M, Vandamma A-M (2009) The phylogenetic handbook: a practical approach to phylogenetics analysis and hypothesis testing, 2nd edn. Cambridge University Press
Nei M, Kumar S (2000) Molecular evolution and phylogenetics. Oxford University Press
Wiens JJ (2000) Phylogenetic analysis of morphological data. In: Smithsonian series in comparative evolutionary biology

Origins of Tree Building: Why Build Evolutionary Trees? 2

Abstract

This chapter traces the history and rationale of evolutionary tree building, beginning with Linnaean taxonomy and the shift to Darwinian systematics. Competing schools of systematics—phenetics (numerical taxonomy) and cladistics (phylogenetic systematics)—are compared, showing how reliance on overall similarity vs. shared derived characters leads to different phylogenetic hypotheses. In this regard, homology and homoplasy are compared and contrasted. Furthermore, various kinds of characters and their use in inferring evolutionary relationships, such as synapomorphies, symplesiomorphies and autapomorphies, are discussed. The chapter argues that phylogenies provide more objective, reproducible frameworks than traditional taxonomies, making them essential for modern biology. The chapter ends by comparing and contrasting the use of molecular data vs. morphological data in tree building.

Keywords

Phenetics · Cladistics · Synapomorphy · Homology · Homoplasy

Biological diversity needs to be discovered, organised and catalogued, i.e. classified. This exercise of classifying biodiversity is of immense importance because we need to know what's out there and how they are related to each other to better understand the underlying biological processes that have generated them. This information, in turn, can be used to our benefit, such as to develop a cure for various diseases or to manage and control various pests, and for efficient use of economically important species (domesticated plants and animals). Moreover, this information is central to the management and conservation of our biological heritage. For example, what do we conserve if we don't know what's out there? Thus, it is very important to characterise the living world and determine their evolutionary relationships.

K. P. Karanth, *Molecular Phylogenetics*, https://doi.org/10.1007/978-981-95-9147-3_2

Classification

The art of classifying organisms was first formalised by Carolus Linnaeus in the eighteenth century. He is credited with the establishment of the branch of science known as 'Linnaean taxonomy'. Taxonomy involves naming, ranking and classifying organisms (also called hierarchical taxonomy). Taxonomy is based on shared similarities, i.e. species that share certain morphological characters are placed in the same genus; genera that share certain characters are placed in the same family and so on. After Darwin published 'On the origin of species' in 1859, it became apparent that classification of biological entities should be based on evolutionary relationships. Thus was born the area of 'Systematics' which deals with the classification of organisms and understanding their evolutionary relationships. Nevertheless, it must be noted that information regarding evolutionary relationships is implicit in taxonomy. This is because overall similarities often suggest evolutionary relationships; in other words, closely related species tend to be morphologically similar. Thus, taxonomy and systematics are the same in that both deal with the classification of the living world, but in the case of the latter, information regarding evolutionary relationships is explicitly incorporated.

Nevertheless, classification is often based on informed opinions of taxonomists wherein relationships between species are based on assertions by experts of various taxonomic groups. Thus, these classifications tend to be subjective and lack objective criteria. This lack of a unifying or standardised classification method or formalised procedure leads to conflicting taxonomies. To resolve these issues, two major schools of systematic classification of organisms arose.

Two Major Schools of Systematics

Phenetics: Classification based on overall similarity in as many characters as possible, usually without weighting (also called numerical taxonomy). In this approach, the various taxa are first 'typed' for a slew of morphological characters. Figure 2.1 has a hypothetical list of characters (characters 1–8). A given character can attain different states in various species. For example, the presence of a crest in birds can be taken as a character (character 4). Some species might lack a crest; in such cases, this character can be coded as '0' for the absence of the character (species C). The crest length might be medium in some species and tall in others; accordingly, they

	1	2	3	4	5	6	7	8
Species A	0	1	1	1	0	1	2	0
Species B	0	1	1	1	0	0	1	0
Species C	0	0	1	0	0	0	1	1
Species D	1	0	2	2	0	0	0	1
Species E	1	0	2	2	1	1	2	1

Fig. 2.1 Species A–E typed for characters 1–8. The numbers 0, 1 and 2 within the box represent various character states. Each column represents a character; character 4 has three states: 0, 1 and 2

can be coded as '1' and '2', respectively. Thus, the character crest has three states 0, 1 and 2. Similarly, other characters can be typed.

These characters are then used to derive an overall pairwise similarity or dissimilarity index (Fig. 2.2). Pairwise dissimilarity is calculated by simply counting the number of characters that differ between the two species.

This index is used to cluster taxa based on dissimilarity, i.e. taxa that exhibit least dissimilarity are placed together in a tree. A tree here refers to a visual representation of relationships between species (Fig. 2.3). Typically, a clustering algorithm such as unweighted pair group method with arithmetic mean (UPGMA) (Sokal and Sneath 1963) is used (this method is discussed later). The output of a phenetic analysis is called a phenogram.

Phenogram: A branching diagram that links entities by estimates of overall dissimilarity

Cladistics: The cladistic approach also uses similarity to group taxa together but does not use all the characters. It distinguishes between homology (similarity due to common ancestry) and homoplasy (convergence) and uses only evolutionary homologies for grouping taxa (discussed in next section). This approach uses the principle of parsimony to derive an optimal tree (optimal cladogram). Parsimony is the principle of choosing the simplest explanation to resolve a problem. An optimal cladogram is a tree requiring the minimum number of steps to explain the character distribution (parsimony is discussed later). The output of a cladistic analysis is called a cladogram (Fig. 2.4).

Fig. 2.2 A matrix showing the pairwise difference (dissimilarity index) between species. This matrix was derived from Fig. 2.1

	A	B	C	D	E
SpeciesA	0				
SpeciesB	2	0			
SpeciesC	5	3	0		
SpeciesD	7	6	4	0	
SpeciesE	6	8	6	3	0

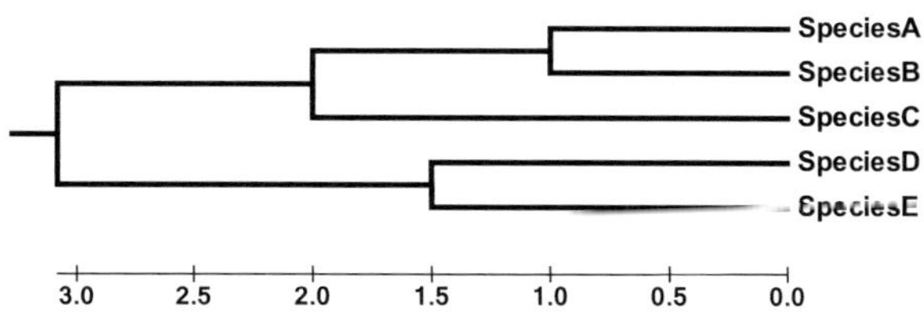

Fig. 2.3 A phenogram showing relationships between species A and E—branch lengths show the dissimilarity between taxa

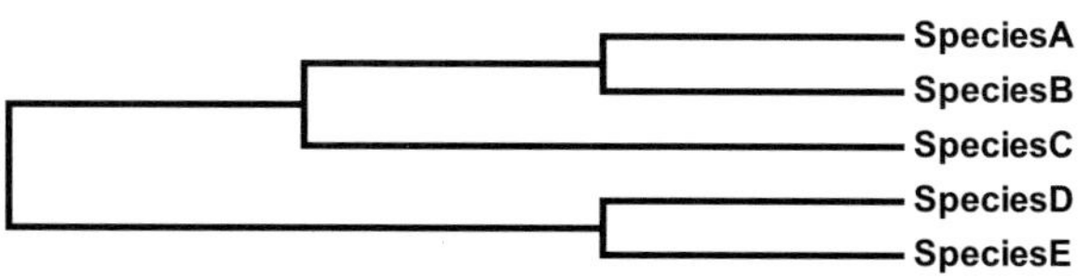

Fig. 2.4 Cladogram showing relationships between species A and E

Cladogram: A tree that depicts inferred historical branching (evolutionary) relationships among species. Unless otherwise stated, the branch lengths in a cladogram are not informative; only the branching order is significant.

Homology Versus Homoplasy

To understand cladistics, we have to first understand the difference between homology and homoplasy. Homology is similarity due to common ancestry. For example, whales, bats and terrestrial mammals (like humans) share the character—the presence of mammary glands (milk-producing glands). This similarity is due to their shared ancestry, in that the common ancestor of whales, bats and terrestrial mammals had mammary glands. These glands did not evolve independently in these taxa. Such similarities are also referred to as 'evolutionary homologies'. Cladistics is a method of classifying organisms (taxa) into groups based on evolutionary homologies (Hennig 1966). In particular, cladistics uses a subset of similarity attributable to shared-derived characters or synapomorphies to group taxa. Synapomorphies (shared-derived characters) are homologous characters (evolutionary homologies) inherited from the most recent common ancestor. To better understand this term, let's map the characters in Fig. 2.1 onto the cladogram in Fig. 2.4. For example, character 1 is present (1) in species D and E but absent (0) in the rest. Assuming that the absence of the character (0) is the ancestral state, we can then come up with the character state for all the internal nodes using the principle of parsimony. This is depicted in Fig. 2.5a where character state '1' in species D and E is a synapomorphy because this character has been inherited from their immediate common ancestor, which also has the same state for this character. The characters 1–4 in Fig. 2.1 are synapomorphies, and these characters have been used to derive the cladogram (Fig. 2.4). In Fig. 2.5, these four characters have been mapped onto the evolutionary tree. The remaining characters (5, 6, 7, 8) are uninformative in a

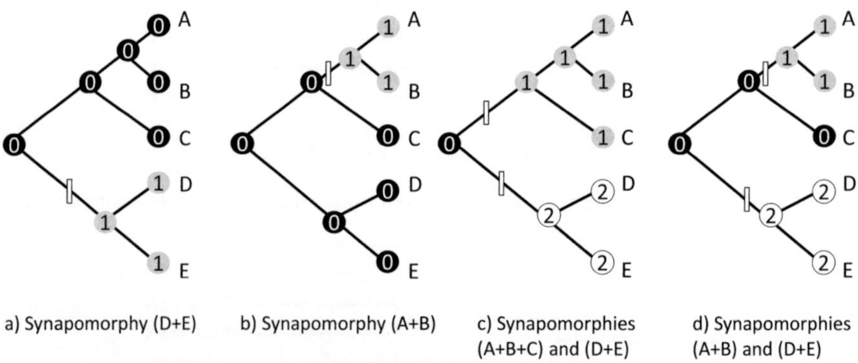

Fig. 2.5 Examples of synapomorphies that unite various species clusters. Open bars represent a change in the character state (character transition) along a lineage

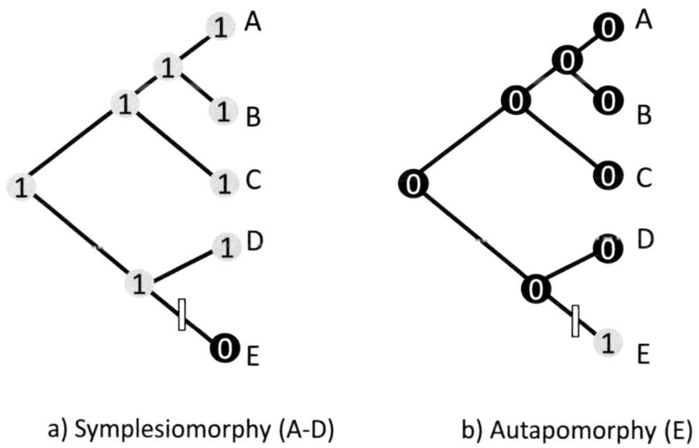

Fig. 2.6 (**a**) Symplesiomorphy and (**b**) autapomorphy

cladistic framework and are classified as symplesiomorphies, autapomorphies and homoplasies (see below).

Symplesiomorphies (shared primitive/ancestral characters) are characters inherited from a remote ancestor as opposed to the most recent common ancestor, as in the case of synapomorphy. This character might have been lost in one or more of the descendants. For example, in Fig. 2.6a, the character coded as '1' is shared by species A–D. This character has been inherited unchanged (except in species E) from the remote ancestor (the root of the tree), and the ancestral state in this case is the presence of the character (1). However, this character was lost along the lineage leading to species E. Thus, the similarity among species A–D for this character is due to symplesiomorphy or shared ancestral characters. The character state '1' in this case is called a plesiomorphic state and is presumed to have been present in the ancestral morphotype. This character state is plesiomorphic in species A–D.

Autapomorphies are characters unique to one species or uniquely derived. In Fig. 2.6b, the character state '1' in species E is an autapomorphy as it is unique to this species. Here, the lack of a particular character in species A–D results in these species appearing to be more similar.

Homoplasy: Similarity in character state due to independent evolution through reversals, convergence or parallel evolution. In the case of reversals, a given character reverts back to its ancestral state. Take the example illustrated in Fig. 2.7a. Here we see a change in the character state (character transition) from 0 to 1 on the lineage leading to the common ancestor of D and E. If this character state were inherited as is in the daughter species (D and E), we would call it a synapomorphy. However, in this case, the character state has reverted back to the ancestral state (0) on the lineage leading to species D. Thus, species D is now more similar to species A, B and C than to its sister species E. In the case of parallel evolution, unrelated taxa (A and E in Fig. 2.7b) independently acquire the same character state. Here the character transition from 0 to 1 has occurred from the same ancestral state (0) along the lineages leading to A and E. Whereas in the case of convergence, unrelated taxa

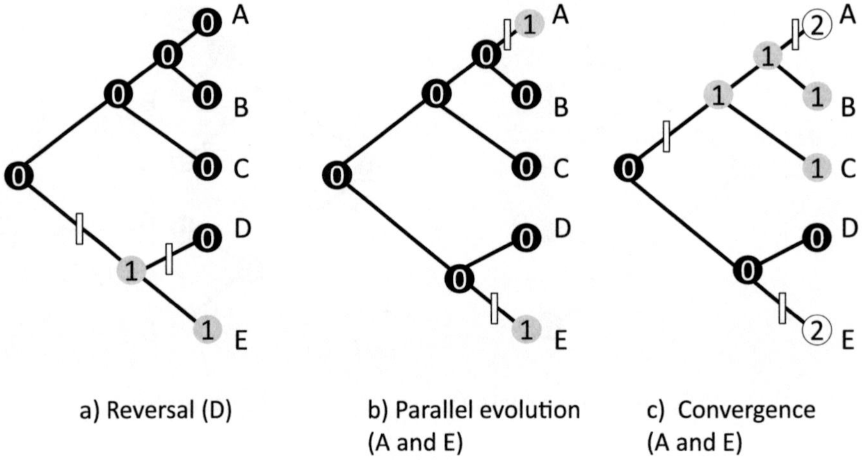

Fig. 2.7 Illustration of homoplasy (reversal, parallel evolution and convergence)

independently acquire the same character from a different ancestral state (A and E in Fig. 2.7c). Here the character transition is from 1 to 2 on the lineage leading to A, whereas it is 0 to 2 on the lineage leading to E. The wing of a bat and the wing of a bird are good examples of homoplasy or, more specifically, similarity due to convergence. Such characters, when used in a phylogenetic analysis, can generate an incorrect phylogeny; for example, bats (mammals) might be placed with birds rather than with other mammals. (*Also see* the chapter on homologous genes).

In cladistics, only synapomorphies contribute towards determining relationships between species; all other characters (homoplasies, apomorphies and symplesiomorphies) are uninformative. A priori, how do we know which of the characters represent synapomorphies?

In cladistic analysis, we use the principle of parsimony to pick the tree that best explains the character distribution (optimal tree). In parsimony, we have to first build all possible trees for a given number of taxa. Characters are then mapped onto these trees as was done in Figs. 2.5, 2.6 and 2.7 for one of the trees. Each time a character appears or is lost in a given tree is considered one evolutionary event. We then add up these events to derive the total tree length for each tree. The tree with the lowest tree length is the most parsimonious tree (optimal tree). Characters that are consistent with this tree are synopomorphies, while others are categorised as homoplasies, apomorphies and symplesiomorphies (*see* Chap. 5 for details regarding parsimony).

Now let us add homoplasies to our original data matrix shown in Fig. 2.1. This gives us Fig. 2.8, where characters 9–12 are homoplasies, mainly characters that exhibit convergent evolution (10 and 11) or reversal (9 and 12). Will these additional characters change the tree topology?

When this new character matrix, shown in Fig. 2.8, is subjected to cladistic analysis, the cladogram generated is the same as in Fig. 2.4. In this case, adding

Fig. 2.8 Characters 9–12 are homoplasies that have been added to the original matrix shown in Fig. 2.1

	1	2	3	4	5	6	7	8	9	10	11	12
Species A	0	1	1	1	0	1	2	0	0	1	1	0
Species B	0	1	1	1	0	0	1	0	0	0	0	0
Species C	0	0	1	0	0	0	1	1	0	0	0	0
Species D	1	0	2	2	0	0	0	1	0	0	0	0
Species E	1	0	2	2	1	1	2	1	1	1	1	1

Fig. 2.9 Distance matrix based on character matrix shown in Fig. 2.8

	A	B	C	D	E
SpeciesA	0				
SpeciesB	4	0			
SpeciesC	7	3	0		
SpeciesD	9	6	4	0	
SpeciesE	8	12	10	7	0

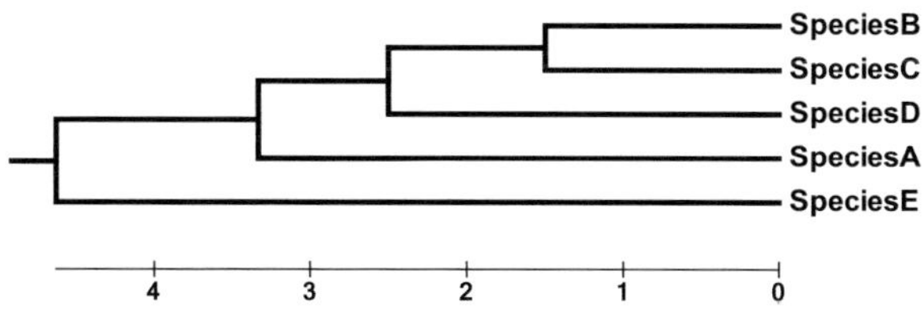

Fig. 2.10 UPGMA tree based on distance matrix shown in Fig. 2.9

homoplasies has not altered the cladogram, and we are still retrieving the right tree. This is because a cladistic analysis uses homologies (synapomorphies) and not homoplasies to retrieve the tree. This is not the case with phenetic analysis. Figure 2.9 is the new distance matrix derived after adding the homoplasies. The UPGMA tree (Fig. 2.10) generated from this distance matrix is very different from the earlier tree that was based on data with fewer homoplasies (Fig. 2.3). In this tree, species E is sister to all other species and also shows a very high distance from other species. This high distance is due to homoplasy (*see* character state for characters 5–12 in species E, Fig. 2.8).

Thus, it is often argued that a cladistic analysis is superior as it distinguishes between homology and homoplasy and uses only synapomorphies (shared derived homologies) to build trees. Whereas the phenetic method uses overall similarity, that is based on both homologous as well as homoplastic characters, to generate a tree. Therefore, 'cladists' would argue that a cladogram is the true evolutionary tree or phylogeny as it uses evolutionary homologies to infer relationships. Whereas a phenogram is usually a good indicator of evolutionary relationships, it is, strictly speaking, not a phylogeny. Nevertheless, it must be pointed out that as we add more and more homoplasies to the dataset, the chances of cladistic analysis generating an erroneous tree increase. This is particularly true in the case of datasets with characters that exhibit convergence and parallel evolution. The principle of parsimony used in the cladistic analysis is not able to distinguish convergence and parallel

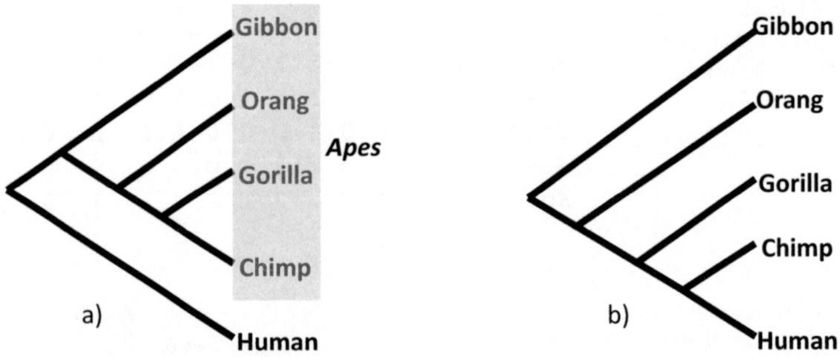

Fig. 2.11 (**a**) Human-ape relationship inferred in the past based on morphological characters and (**b**) current widely accepted relationship based on molecular data

evolution from true synapomorphies. Thus, as long as true synapomorphies outnumber homoplasies, cladistic analysis will generate the right tree.

Other class of characters that are problematic in phenetic analysis are symplesiomorphies and autapomorphies. These characters, when used in an analysis, can generate erroneous trees.

Take the example of the human-ape phylogeny (Fig. 2.11). Apes appear to be more similar to each other in their morphology as compared to humans. Thus, it was widely believed that apes were more closely related to each other and hence were considered a natural group. Consequently, they were classified in a separate family distinct from humans (Fig. 2.11a). It was only since the 1960s that genetic and molecular studies began to support the current phylogeny (Fig. 2.11b). Today, we know that humans are nested deep within the so-called apes and that chimps (an ape) are more closely related to humans than they are to any other ape. The overall similarity among apes is due to autapomorphies along the human lineage (Fig. 2.12) and symplesiomorphies (shared ancestral characters) among the apes (Fig. 2.13). If these characters are used to build an evolutionary tree, then all apes will group together and humans will be distinct (Fig. 2.11a). Therefore, overall similarity can be deceptive with respect to building an evolutionary tree. Autapomorphies on the human lineage include obligate bipedalism, larger cranial capacity, large brain, complex language, tool use, cognition, and numerous other skeletal characters.

Why Build Evolutionary Trees (Phylogenies)?

A phylogeny is an evolutionary tree that shows how different species are related to each other. One of the fundamental applications of phylogenies is in classification. Phylogenies help us systematically classify organisms in a manner that is consistent with their evolutionary relationships. Such a classification is less prone to errors and individual biases. Phylogenies are an important component in the field of historical biogeography (vicariance vs. dispersal and phylogeography).

	1	2	3	4
Gibbon	0	0	0	0
Orang	0	0	0	0
Gorilla	0	0	0	0
Chimp	0	0	0	0
Human	1	1	1	1

a)

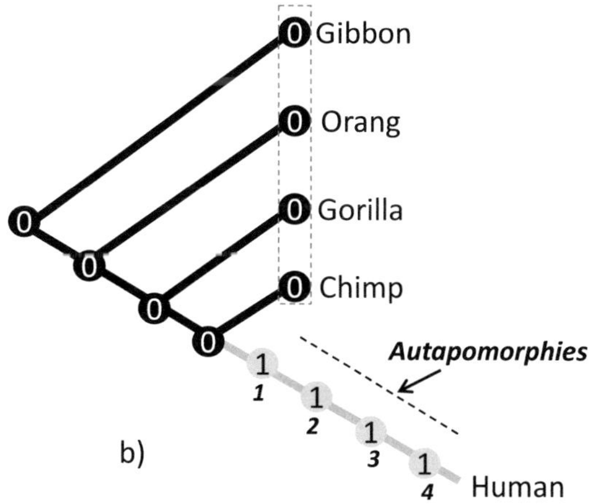

b)

Fig. 2.12 (**a**) Characters 1–4 are autapomorphies in humans. (**b**) These character transitions have occurred on the lineage leading to humans after humans and chimps diverged from each other

	5	6	7	8
Gibbon	1	1	1	1
Orang	1	1	1	1
Gorilla	1	1	1	1
Chimp	1	1	1	1
Human	0	0	0	0

a)

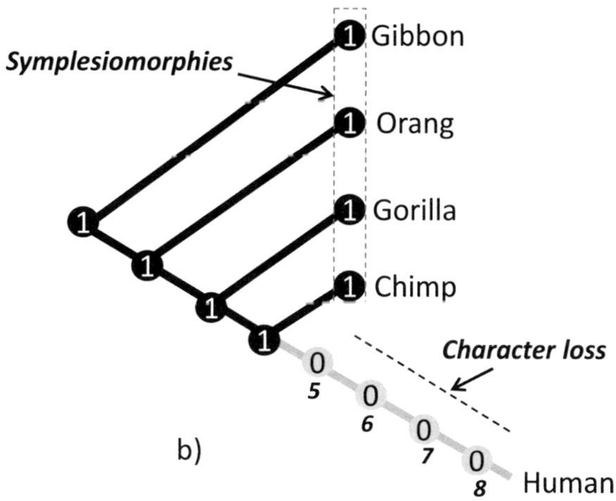

b)

Fig. 2.13 (**a**) Characters 5–8 are symplesiomorphies among apes. (**b**) These characters have been lost on the lineage leading to humans after humans and chimps diverged from each other

Furthermore, they are useful in studying the evolution of species (budding speciation, sympatric vs. allopatric) and in identifying cryptic species. The study of evolution of behavioural and morphological characters in a phylogenetic context (character evolution) is also an exciting application of phylogenetics. Gene evolution too can be better studied in a phylogenetic framework (molecular evolution). These applications are discussed in Chap. 10. Phylogenies can be built using molecular or morphological data. So far, we have discussed how phylogenies are built using morphological data, but in recent times, molecular data are increasingly being used. There are some major advantages using molecular data. Here we compare morphological data with molecular data.

Morphological Data

One of the major issues with morphological data is the lack of variable characters to generate robust phylogenies. Typically, in morphology-based studies, the number of characters is limited (usually less than 100), and often these characters do not exhibit much variation across taxa. Thus, phylogenetic analysis based on morphological characters often generates poorly supported trees (*see* section on bootstrapping, Chap. 9). Another issue with morphological data is deciding the kind of data to be recorded. For example, one could use skeletal characters, coat colour or external morphological characters such as scale counts or shape. Depending on the characters used and how they are coded, one might get a different tree. For example, in Fig. 2.14, depending on the character used and how it is coded, one can generate different phylogenetic hypotheses. If 'shape' is used, then species 2 and 3 are sister taxa, whereas if 'colour' is used, then species 1 and 2 are sister taxa.

One could get around this issue by including all the characters irrespective of what type they are. However, this could often be problematic as some characters are discrete (e.g. shape) while others are continuous (e.g. length) or some are binary (present or absent) while others are multistate, etc.

One potentially limiting factor in the use of morphological data in generating phylogeny is the requirement of a trained taxonomist to code the characters. For example, if an evolutionary biologist is interested in the phylogeny of a particular group of butterflies, then she has to either get trained in butterfly taxonomy or take the help of a lepidopterist. Often, for certain groups, there are very few expert taxonomists, thereby seriously limiting phylogenetic studies. More importantly, data

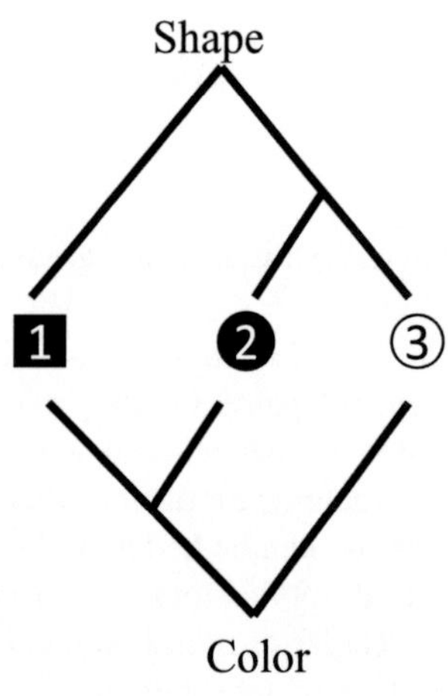

Fig. 2.14 Example of incongruence in tree topology due to different coding schemes

acquisition techniques are not identical across taxa. For example, characters used to generate the phylogeny of fishes are very different from those used for insects. Nevertheless, acquisition of morphological data is usually inexpensive, and often all one needs is a good microscope.

Molecular Data

The major advantage of molecular data is access to a large number of variable characters. Depending on the availability of time and resources, one can either sequence a few genes or the complete genome of an organism. Typically, molecular data also exhibit higher levels of variation. Secondly, it is easier to code molecular data, and there is no ambiguity in how it should be coded. For example, in the case of DNA sequence data, there are four possible character states: A, T, G and C. Irrespective of the taxonomic group under consideration, the coding of these nucleotides remains the same. Molecular data are also comparable across taxonomic groups. For example, cytochrome *b* sequences can be compared between distantly related groups such as insects and vertebrates. However, the same is not true in the case of morphological data; the wing of a bird cannot be compared with that of an insect.

Most importantly, the data acquisition technique is identical across taxa, i.e. irrespective of the group under consideration, the molecular tools used are the same. For every group, we start with DNA extraction from a tissue sample, followed by PCR amplification of the selected markers and sequencing. Thus, any biology student can be trained to undertake molecular work, and they can potentially work on any group. Nevertheless, one still needs a taxonomist for accurate identification of the specimens. However, the actual coding of the data can be undertaken by anyone, and there is no requirement for a taxonomist. The major disadvantage of molecular data is that generating sequence data can be very expensive. Typically, one needs a fairly well-equipped molecular lab, and the consumables used for generating sequence data can be expensive. Additionally, phylogenetic analyses based on molecular data also face some of the issues of morphological data, for example, what gene or marker to use? Do we combine data from multiple markers or analyse them separately? Some of these issues are addressed in subsequent chapters.

This does not mean that traditional morphology-based taxonomy should be abandoned. Let us not forget that most of the described species on the planet have been identified and classified based on morphological characters. Thus, morphology-based taxonomy has played a central role in our understanding of biology in general, and biodiversity in particular. Moreover, a molecular phylogeneticist needs a taxonomist for species identification. By and large, morphology-based classification schemes have also been quite useful in inferring evolutionary relationships. But on numerous occasions (both at lower-level and higher-level taxonomy), there has been poor concordance between morphology-based classification and phylogeny (e.g. human-ape relationship). Thus, morphological characters are ideal for species identification, description and delimitation, but morphology-based taxonomy might not be a good estimate of evolutionary relationship. Molecular data are better suited for inferring phylogenies or evolutionary relationships. The need of the hour is active collaboration between taxonomists and molecular phylogeneticists and the need to develop a synergistic association between these two fields. Taxonomists should not see molecular phylogeneticists as competitors because phylogeny complements taxonomy.

References

Hennig W (1966) Phylogenetic systematics. University of Illinois Press, Urbana

Sokal RR, Sneath PHA (1963) Principles of numerical taxonomy. W.H. Freeman & Co., New York

Methods of Tree Building Using DNA Sequence Data-I: Distance Approach

3

Abstract

This chapter introduces distance-based tree-building methods. It begins with the calculation of pairwise genetic distances from aligned DNA sequences and shows how matrices are generated. The unweighted pair group method with arithmetic mean (UPGMA) is presented as a clustering algorithm that assumes a molecular clock, with limitations highlighted. To account for rate variation across lineages, the Fitch-Margoliash method is discussed. The neighbour-joining (NJ) method is introduced as a more flexible alternative that accommodates rate variation across lineages. Optimality criteria-based approaches, including minimum evolution and least squares, are explained as improvements over simple clustering, as they evaluate multiple topologies.

Keywords

UPGMA · NJ · Optimality criteria · Branch length · Rate variation

In distance-based tree-building methods, we first calculate the pairwise genetic distance between species, and these distances are then used to build the tree. In the example shown in Fig. 3.1, a variable region of a gene consisting of nine sites (nucleotides) has been sequenced from four species, and these sequences have been aligned. Among the nine sites, the second and sixth sites are identical across taxa; such sites are uninformative as they do not provide information on how species might be related.

Next, we determine the differences in the sequences between all possible pairs of species. This is done by simply counting the number of differences in the nucleotides for each site between each species pair as shown in Table 3.1.

This pairwise distance, also called genetic distance, is used to generate a distance tree using various methods.

K. P. Karanth, *Molecular Phylogenetics*, https://doi.org/10.1007/978-981-95-9147-3_3

Fig. 3.1 An alignment consisting of sequences from four species. Numbers indicate the nine sites in the sequence

```
         123456789
Human    GTCACATGT
Chimp    CTCACATCT
Gorilla  CTCCCATTC
Orang    ATACGAGCC
```

Table 3.1 A distance matrix showing pairwise sequence difference among four species

	H	C	G	O
Human (H)	0			
Chimp (C)	2	0		
Gorilla (G)	4	3	0	
Orang (O)	7	6	5	0

Unweighted Paired Group Method with Arithmetic Mean (UPGMA)

In the UPGMA method, we start with the species pair with the lowest genetic distance. In our example, it is between human and chimp (distance of 2). These two species are placed together and the node is placed at a distance of one unit from each species, i.e. overall distance divided by the number of species (Fig. 3.2).

Next, human-chimp (HC) is considered a taxon, and the distance between HC and the remaining species is calculated. The distance between HC and gorilla (G) is given as the distance between H and G plus the distance between chimp (C) and G, divided by two.

Thus, we get

$$Dis_{HC-G} = \left(Dis_{H-C} + Dis_{C-G}\right)/2 = \left(4+3\right)/2 = 3.5$$

Similarly, we calculate the distance between HC and orang (O):

$$Dis_{HC-O} = \left(Dis_{H-O} + Dis_{C-O}\right)/2 = \left(7+6\right)/2 = 6.5$$

We now have a new distance matrix, wherein H and C have been combined into a taxon called human-chimp (Table 3.2).

In this matrix, the smallest distance is between HC and G (3.5); these taxa are placed in a cluster, and the node representing the common ancestor of H, C and G is placed at 3.5/2 = 1.75 (Fig. 3.3).

The last member of the tree is orang (O). Thus, next we calculate the distance between (HCG) and O.

$Dis_{HCG-O} = (Dis_{H-O} + Dis_{C-O} + Dis_{G-O})/3 = (7 + 6 + 5)/3 = 6$, the node representing the common ancestor of H, C, G and O is placed at 6/2 = 3.

The tree generated by UPGMA method is called a phenogram (Fig. 3.4). A phenogram provides information regarding species relationship as well as the genetic distance between them. Thus, in a phenogram, the branch lengths are also

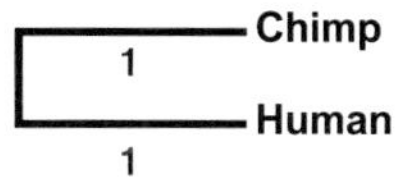

Fig. 3.2 Human and chimp clustered together with one change along each lineage

Table 3.2 Revised distance matrix where human and chimp are considered a taxon

	HC	G	O
Human-chimp (HC)	0		
Gorilla (G)	3.5	0	
Orang (O)	6.5	5	0

Fig. 3.3 Human, chimp and gorilla clustered together with 1.75 changes along each lineage

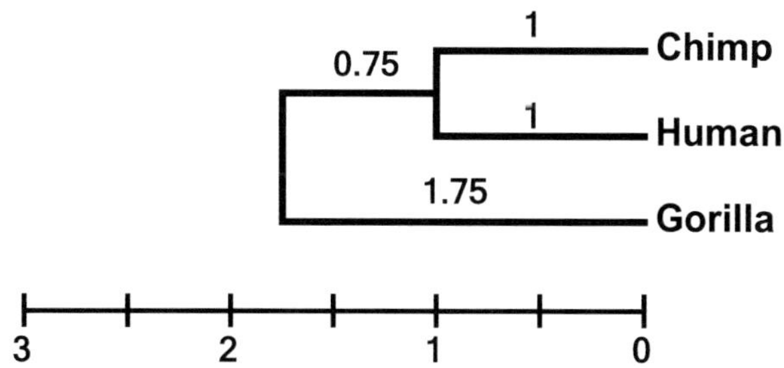

Fig. 3.4 A phenogram showing the relationship between human, chimp, gorilla and orang

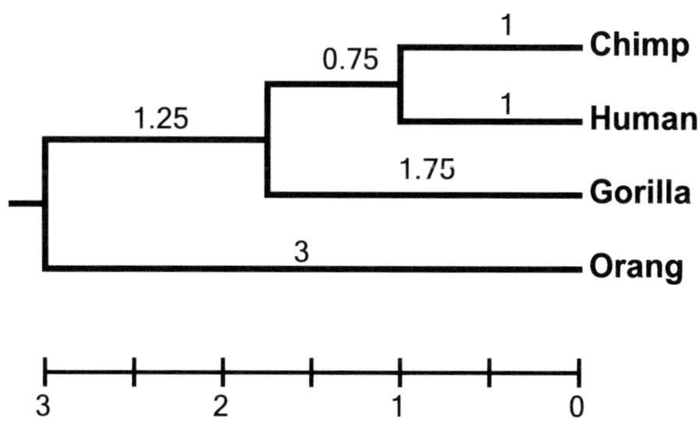

informative as they provide information regarding the number of changes along each lineage. In this regard, it is similar to a phylogram (*see* Chap. 1). The UPGMA phenogram has one interesting feature, in that all the extant species are at the same distance from the root (Fig. 3.4), and by default, the tree is rooted. For example, the distance from human to the root is the sum of the branches connecting human with the root, which is 3 (1 + 0.75 + 1.25). It is the same distance from the root to all other species. This means the number of changes that have accumulated along the lineages (branches) leading to the different species is the same. Thus, UPGMA method assumes equal rates of molecular evolution along different lineages. We call this condition a molecular clock, i.e. the chosen marker is evolving in a clock-like manner and at the same rate along different lineages.

Table 3.3 Distance matrix showing pairwise tree distance among the four species

	H	C	G	O
Human (H)	0			
Chimp (C)	2	0		
Gorilla (G)	3.5	3.5	0	
Orang (O)	6	6	6	0

The molecular clock assumption of UPGMA method generates a phenogram in which the distances between taxa on the tree (tree distance) do not correspond to the observed distances. The observed distance is shown in Table 3.1, and the tree distances for the same species pairs are shown in Table 3.3. Tree distances are calculated by adding the branch lengths between species in the tree. For example, the tree distance between human and gorilla is 3.5 (1 + 0.75 + 1.75), whereas the observed distance is 4.

The UPGMA method generates average distances among taxa and uses these modified distances to build the tree. For example, the distance between gorilla and human as well as between gorilla and chimp is 3.5, which is an average distance between human-chimp and gorilla (4 + 3/2). Similarly, the distance between orang and all other species is an average of distances between orang and other species (7 + 6 + 5/3 = 6).

The advantage of UPGMA is that it needs very little computational time and generates a rooted tree. However, it assumes molecular clock and is very sensitive to violations of this assumption. The assumption of strict clock is very unrealistic; thus, in most cases, UPGMA might generate an erroneous relationship. Given that there are other superior clustering algorithms (such as NJ), there is no need to use UPGMA in phylogenetic analysis. Many statistical software have an option to generate a tree or a dendrogram; usually, the tree-building method used there is UPGMA.

Neighbour-Joining Method (NJ)

As mentioned earlier, the major drawback of the UPGMA method is the molecular clock assumption, which is often unrealistic, because, in nature, substitution rates vary among lineages. The molecular clock assumption results in a phenogram, wherein the tree distance does not correspond to the observed distance. Ideally, the tree-building method should generate an additive tree, i.e. the branch lengths should add up to the observed distances between species. The NJ method sequentially finds pairs of closely related taxa after correcting for the rate variation across lineages, and it also generates an additive tree. The first step in the NJ method is to calculate the rate-corrected distances among taxa.

To do this, we first calculate the total distance (*td*) between any one taxon and all the other taxa. The total distance for human would be the sum of genetic distances between human and all other species in the distance matrix above (Table 3.4).

Table 3.4 A distance matrix showing pairwise sequence difference among four species

	H	C	G	O
Human (H)	0			
Chimp (C)	2	0		
Gorilla (G)	4	3	0	
Orang (O)	7	6	5	0

$$td_H = 2 + 4 + 7 = 13$$

Similarly, we get $td_C = 11$, $td_G = 12$, $td_O = 18$.

Next, we calculate the rate-corrected distances (*rd*) among taxa. The *rd* between H and C is given as: $rd_{HC} = d_{HC} - [(td_H + td_C)/(N - 2)]$, where d_{HC} is the observed distance between H and C from the above distance matrix and N is the total number of taxa.

$$rd_{HC} = 2 - (13 + 11)/2 = -10$$

Similarly, $rd_{HG} = 4 - (13 + 12)/2 = -8.5$, $rd_{HO} = 7 - (13 + 18)/2 = -8.5$, $rd_{CG} = 3 - (11 + 12)/2 = -8.5$, $rd_{CO} = 6 - (11 + 18)/2 = -8.5$, $rd_{GO} = 5 - (12 + 18)/2 = -10$.

Thus, we get the rate-corrected matrix (Table 3.5).

In the rate-corrected distance matrix, the lowest distance is between H and C as well as between G and O. Thus, we can take either of the two pairs for our first node. Let us start by connecting G and O to a node X (Fig. 3.5).

Next, we need to calculate the branch lengths (*b*) XG and XO.

$$b_{XG} = d_{GO}/2 + (td_G - td_O)/2(N - 2) = 5/2 + (12 - 18)/2(2) = 2.5 - 1.5 = 1$$

$$b_{XO} = d_{GO}/2 + (td_O - td_G)/2(N - 2) = 5/2 + (18 - 12)/2(2) = 2.5 + 1.5 = 4$$

Thus, we get the branch lengths shown in Fig. 3.6.

We now have to recalculate the distance (*d*) between the node X and the remaining species, H and C (Table 3.6).

$$d_{HX} = (d_{HG} + d_{HO} - d_{GO})/2 = (4 + 7 - 5)/2 = 3$$

$$d_{CX} = (d_{CG} + d_{CO} - d_{GO})/2 = (3 + 6 - 5)/2 = 2$$

We now repeat all the steps:

$$td_X = 5, td_H = 5, td_C = 4$$

Remember now, $N = 3$

$$rd_{HX} = d_{HX} - (td_H + td_X)/(N - 2) = 3 - 10 = -7$$

$$rd_{CX} = d_{CX} - (td_C + td_X)/(N - 2) = 2 - 9 = -7$$

Table 3.5 Rate-corrected distance matrix

	H	C	G	O
Human (H)	0			
Chimp (C)	–10	0		
Gorilla (G)	–8.5	–8.5	0	
Orang (O)	–8.5	–8.5	–10	0

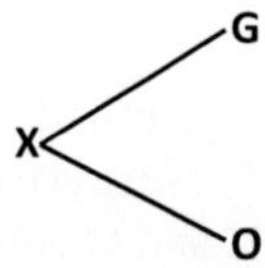

Fig. 3.5 Tree showing the relationship between G and O, with X representing the common ancestor of G and O

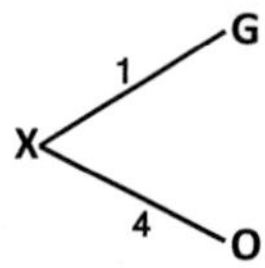

Fig. 3.6 Tree showing branch lengths XG and XO

Table 3.6 A distance matrix showing pairwise distances between XH, XC and CH; distance CH taken from Table 3.5

	X	H	C
X	0		
H	3	0	
C	2	2	0

$$rd_{CH} = d_{CH} - (td_C + td_H)/(N-2) = 2 - 9 = -7$$

We can either take C and H or C and X or H and X to form the next node. Let us take H and X. They are connected to node Y (Fig. 3.7).

Next, we need to calculate the branch lengths (*b*) YH and YX (Fig. 3.8).

$$b_{YH} = \frac{d_{HX}}{2} + (td_H - td_X)/2(N-2) = 3/2 + (5-5)/2 = 1.5$$

$$b_{YX} = \frac{d_{HX}}{2} + (td_X - td_H)/2(N-2) = 3/2 + (5-5)/2 = 1.5$$

Recalculate the distance between Y and the remaining taxon, i.e. C (Fig. 3.9).

$$d_{YC} = (d_{HC} + d_{XC} - d_{HX})/2 = (2+2-3)/2 = 0.5$$

If we scale the branches according to branch lengths, then we get the unrooted phenogram depicted in Fig. 3.10.

We can now root this tree using the midpoint method (outgroup rooting is discussed later). The root is placed at the midpoint (arrow) of the longest tree distance

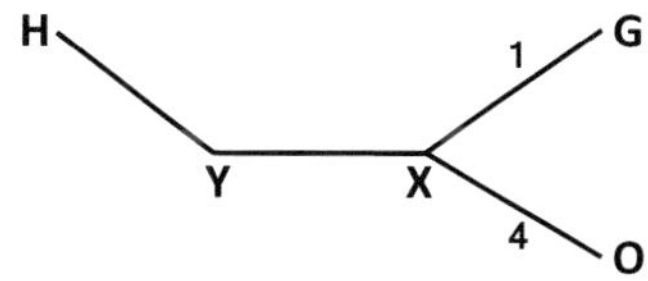

Fig. 3.7 Tree showing relationship between G, O and H

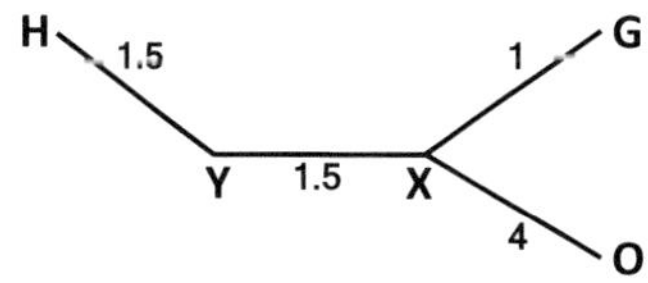

Fig. 3.8 Tree showing branch lengths HY, XY, XG and XO

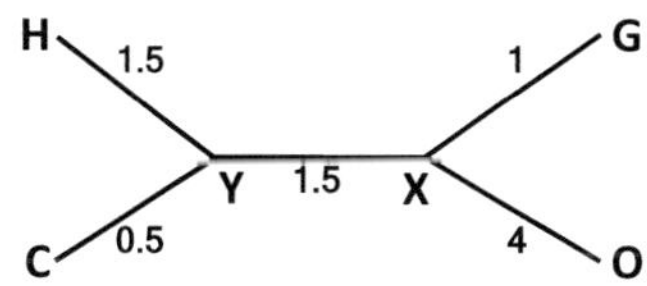

Fig. 3.9 Complete unrooted tree showing branch lengths CY, HY, XY, XG and XO

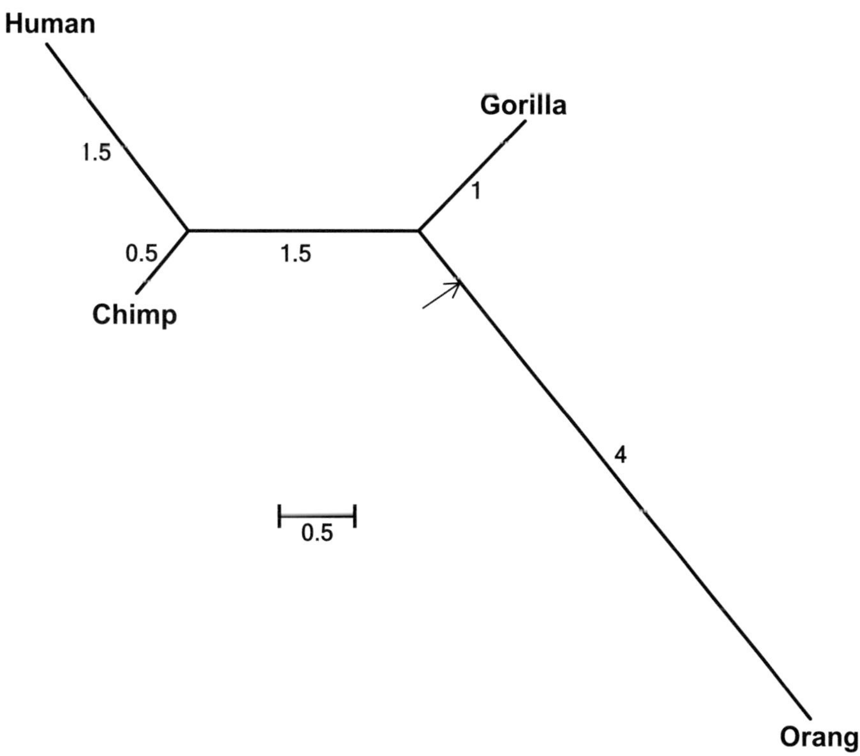

Fig. 3.10 Unrooted phenogram. The arrow represents the midpoint of the longest tree distance between any two taxa (human and orang)

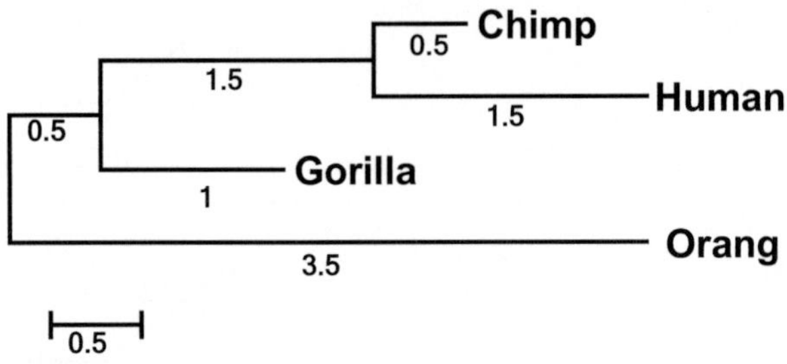

Fig. 3.11 Rooted phenogram

between two taxa. In our case, it is between O and H, and the tree distance between them is 7 (4 + 1.5 + 1.5). The midpoint would be 7/2 = 3.5

The result of midpoint rooting is illustrated in Fig. 3.11.

As is apparent from the rooted NJ phenogram, all the taxa are at different distances from the root, with the exception of H and O, which are at the same distance from the root (3.5). This is due to midpoint rooting. Furthermore, NJ generates an additive tree, i.e. the tree distance between taxa is the same as the observed distance in the distance matrix. These features are due to the fact that the NJ method incorporates rate variation across lineages and does not assume a molecular clock. In this regard, the NJ method is far superior to UPGMA. The NJ method also needs very little time to generate a tree. However, one major drawback of this method is that it does not evaluate all possible trees.

Optimality Criteria-Based Tree-Building Methods

Till now we have discussed tree-building methods (UPGMA and NJ) that generate a tree for a given set of taxa without evaluating other possible tree topologies. In these methods, the branch lengths are calculated as and when the tree is being assembled. However, there are multiple possible trees for a given number of taxa. For example, from the previous exercise, we know that for four taxa, one can come up with three possible unrooted trees (Fig. 3.6, Chap. 1). With respect to great ape-human relationships, these trees would be as shown in Fig. 3.12.

The Question Then Is Which Among These Three Trees Is the True Tree?

UPGMA and NJ methods do not evaluate all these trees. More advanced methods, such as least square and minimum evolution, use an ‘optimality criterion’ to pick the best tree among the three trees. We will discuss these methods as well as optimality criteria in more detail in the next section. For these methods to function, we need the branch lengths of the trees depicted above. Branch lengths, for a given tree,

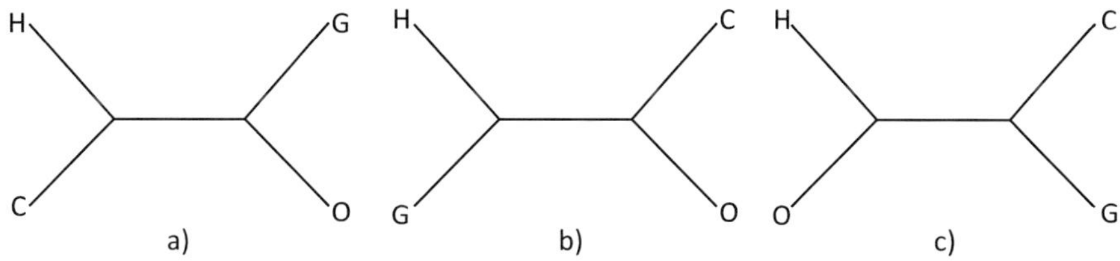

Fig. 3.12 Three possible unrooted trees depicting relationship between apes and humans

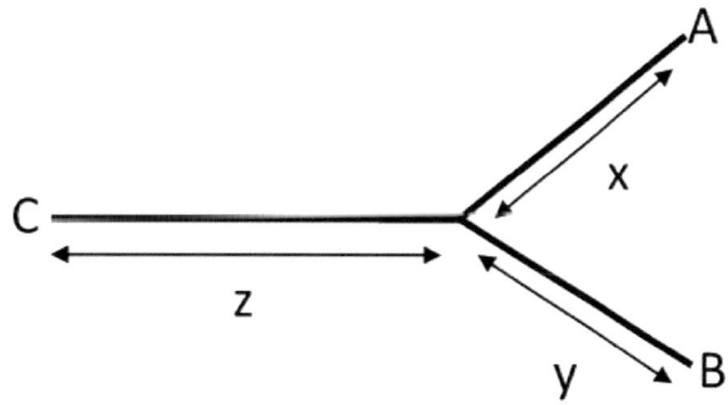

Fig. 3.13 Unrooted tree showing relationship between A, B and C

can be calculated from the distance matrix. Here, we will discuss one of the simplest methods for calculating branch lengths, called the Fitch-Margoliash method (Fitch and Margoliash 1967).

Calculating Branch Lengths of a Given Tree

Let us start with a simple unrooted three-taxa tree depicted in Fig. 3.13. We need to calculate the branch lengths X, Y and Z, which are lineages leading to taxa A, B and C, respectively. From Fig. 3.13, it is apparent that genetic distance (*d*) between A and B (d_{AB}) equals the sum of the branch lengths X and Y.

$$d_{AB} = X + Y$$

Similarly,

$$d_{AC} = X + Z \text{ and } d_{BC} = Y + Z$$

Given the above relationship between branch lengths and genetic distances, X, Y and Z can be calculated using the following formula:

$$X = (d_{AB} + d_{AC} - d_{BC}) / 2$$

$$Y = (d_{BC} + d_{AB} - d_{AC}) / 2$$

$$Z = (d_{BC} + d_{AC} - d_{AB}) / 2$$

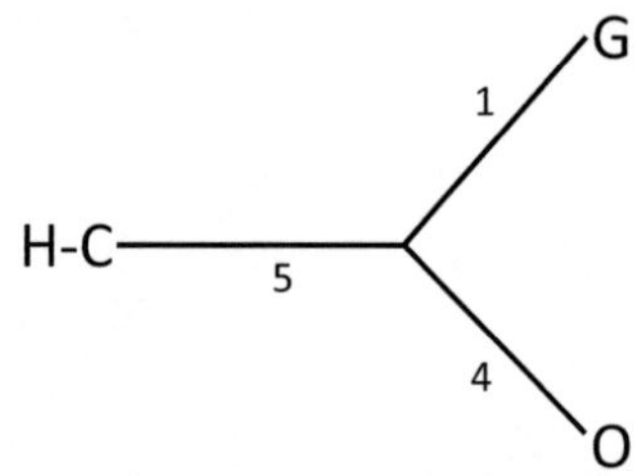

Fig. 3.14 Unrooted tree showing relationship between G (as A), O (as B) and HC (as C)

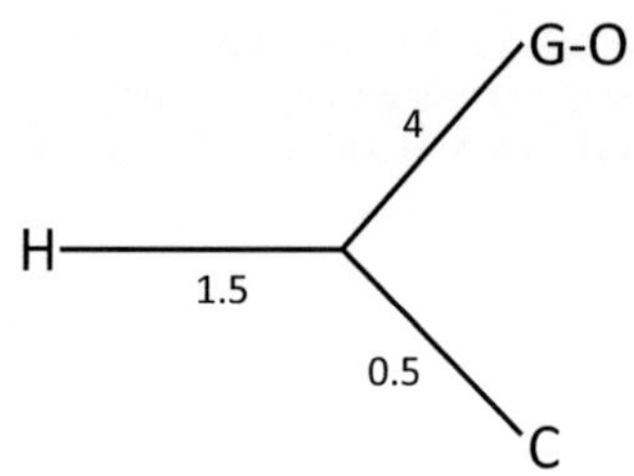

Fig. 3.15 Unrooted tree showing relationship between GO (as A), C (as B) and H (as C)

The values for d_{AB}, d_{AC} and d_{BC} come from the distance matrix (Table 3.4). For four or more taxa, we use the same logic with slight modification. We take gorilla and orang as our taxa A and B (or we could also take H and C). In our example, human and chimp together become taxon C (Fig. 3.14). We recalculate d_{AC} and d_{BC} using the formula given below:

$d_{AB} = d_{GO} = 5$ from the distance matrix

$$d_{AC} = (d_{GH} + d_{GC})/2 = (4+3)/2 = 3.5$$

$$d_{BC} = (d_{OH} + d_{OC})/2 = (7+6)/2 = 6.5$$

Thus, X, Y and Z are

$$X = (d_{AB} + d_{AC} - d_{BC})/2 = (5+3.5-6.5)/2 = 1$$

$$Y = (d_{BC} + d_{AB} - d_{AC})/2 = (6+5-3.5)/2 = 4$$

$$Z = (d_{BC} + d_{AC} - d_{AB})/2 = (6.5+3.5-5)/2 = 5$$

Next, gorilla and orang are combined into a single taxon, and this becomes our new A and chimp (or human) the new B (Fig. 3.15). Next, we need to recalculate the distance between our new A (gorilla-orang) and B (chimp) and C (human).

d_{AB} = Average of the distances between gorilla and chimp, and orang and chimp

$$d_{AB} = (d_{GC} + d_{OC})/2 = (3+6)/2 = 4.5$$

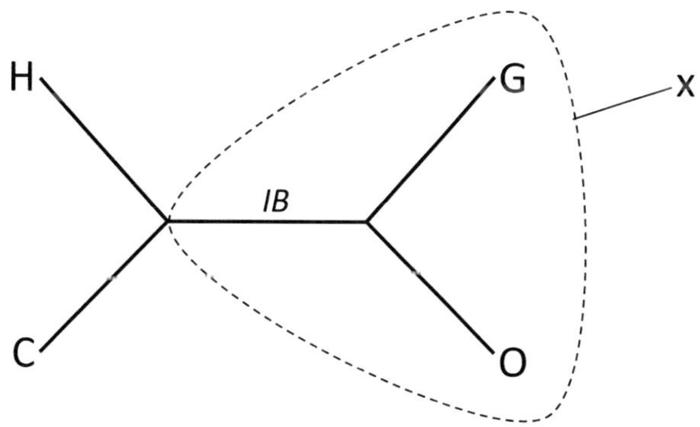

Fig. 3.16 Unrooted tree depicting internal branch (IB)

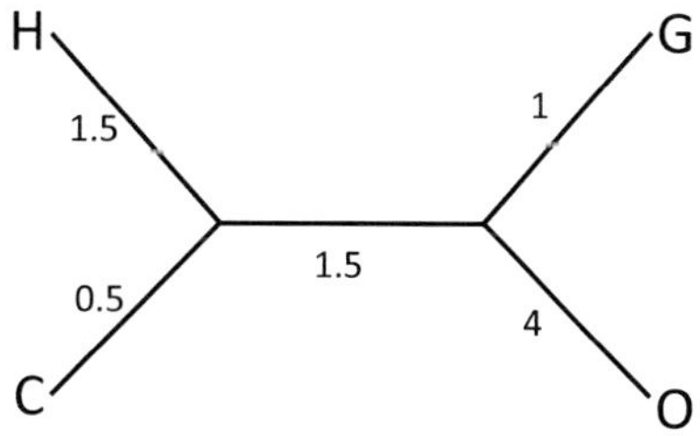

Fig. 3.17 One of the three possible unrooted trees with branch lengths

Similarly,

$d_{AC} = (d_{GH} + d_{OH})/2 = (4 + 7)/2 = 5.5$
$d_{BC} =$ The original distances between chimp and human
$d_{BC} = d_{CH} = 2$ from the distance matrix

$$X = \left(d_{AB} + d_{AC} - d_{BC}\right)/2 = \left(4.5 + 5.5 - 2\right)/2 = 4$$

$$Y = \left(d_{BC} + d_{AB} - d_{AC}\right)/2 = \left(2 + 4.5 - 5.5\right)/2 = 0.5$$

$$Z = \left(d_{BC} + d_{AC} - d_{AB}\right)/2 = \left(2 + 5.5 - 4.5\right)/2 = 1.5$$

In the next step, we expand GO of Fig. 3.15 to give us Fig. 3.16. Here X = branch lengths leading to gorilla and orang + internal branch length (IB)

The branch lengths leading to gorilla and orang are 1 and 4 (Fig. 3.14), respectively, the average branch length is 2.5.

Thus, 4 = 2.5 + IB

$$IB = 1.5$$

This then gives us the following tree with the corresponding branch lengths (Fig. 3.17).

We implement the same procedure on the other two tree topologies (Fig. 3.18).

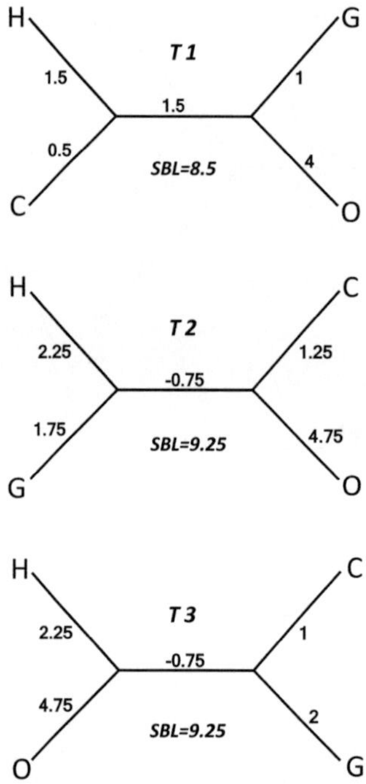

Tree distance

	H	C	G	O
Human	0			
Chimp	2	0		
Gorilla	4	3	0	
Orang	7	6	5	0

	H	C	G	O
Human	0			
Chimp	2.75	0		
Gorilla	4	2.25	0	
Orang	6.25	6	5.75	0

	H	C	G	O
Human	0			
Chimp	2.5	0		
Gorilla	3.5	3	0	
Orang	7	5	6	0

Fig. 3.18 All possible unrooted trees (T1, T2 and T3) for four taxa with branch lengths and sum of branch lengths (SBL). The distance matrix on the right depicts tree distances between taxa

Minimum Evolution (ME)

For a given tree topology, the branch lengths are calculated using the method described in the earlier section. Once the branch lengths are known, these can be added to determine the sum of branch lengths (SBLs) for a given tree. The sum of branch lengths varies depending on the topology. In our example, SBL of tree 1 (T1) is 8.5, whereas for T2 and T3, it is 9.25 (Fig. 3.18). The minimum evolution method tries to find the tree for which the sum of the branch lengths is minimal. Here, all possible tree topologies are first constructed, and the SBLs are calculated for each tree. The ME method then picks the tree with the minimum SBL. Tree 1 has the lowest SBL (8.5), and this is our ME tree.

Least Squares (LS)

Branch lengths can also be used to determine the tree distance (*Td*) between two species. The tree distance between two species is calculated by adding the lengths of the branches connecting them. The tree distance, in turn, can vary depending on

Fig. 3.19 Rooted tree (T1)

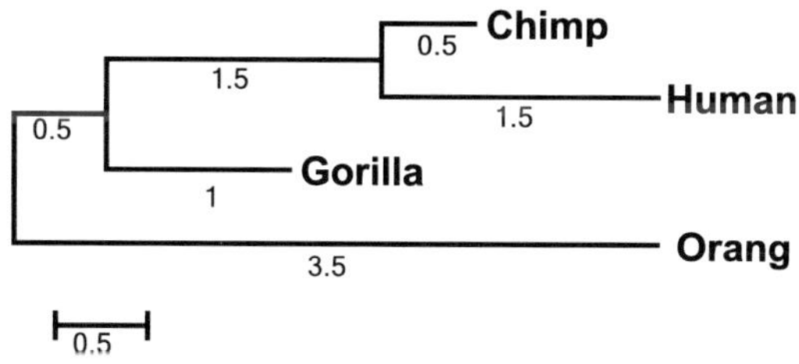

the tree topology. For example, in T1, the *Td* between human and chimp is 2, whereas in T2 and T3, it is 2.75 and 2.5, respectively (Fig. 3.18). The *Td* can be compared with observed distance (*d*) from the distance matrix (Table 3.4). In our example, the *Td* of T1 is identical to *d*; such a tree is called an additive tree (*see* section on NJ). However, there is much discrepancy between *Td* and *d* for trees T2 and T3. This discrepancy varies depending on the tree topology and is used in the least squares method as a measure of how well a particular tree fits the data (*d*). The LS method selects the tree that shows the least discrepancy between *d* and *Td*. In this method, firstly, all possible tree topologies for the given number of taxa are constructed. For each of these trees, the branch lengths are calculated. These branch lengths are, in turn, used to determine *Td* between taxa. The LS method then picks the tree topology for which the difference between tree distance and observed distance is the least.

Least squares score for our four-taxa tree is calculated using the following formula:

$$LS = \left(d_{HC} - Td_{HC}\right)^2 + \left(d_{HG} - Td_{HG}\right)^2 + \left(d_{HO} - Td_{HO}\right)^2 + \left(d_{CG} - Td_{CG}\right)^2 + \left(d_{CO} - Td_{CO}\right)^2 + \left(d_{GO} - Td_{GO}\right)^2$$

$$\text{T1}\,LS = (2-2)^2 + (4-4)^2 + (7-7)^2 + (3-3)^2 + (6-6)^2 + (5-5)^2 = 0$$

$$\text{T2}\,LS = (2-2.75)^2 + (4-4)^2 + (7-6.25)^2 + (3-2.25)^2 + (6-6)^2 + (5-5.75)^2 = 2.25$$

$$\text{T3}\,LS = (2-2.5)^2 + (4-3.5)^2 + (7-7)^2 + (3-3)^2 + (6-5)^2 + (5-6)^2 = 2.5$$

In our example, the LS tree is T1. Thus, both ME and LS methods have picked the same tree, and this tree can then be rooted using midpoint rooting described earlier to give the following tree (Fig. 3.19):

The ME and LS methods use SBL and LS score, respectively, to pick the best tree. These methods attempt to optimise the value of the chosen criteria and therefore are referred to as optimality criteria-based methods. Thus, distance-based methods can be further categorised into methods that use an algorithm (algorithmic approach) to generate a single tree, such as NJ and UPGMA, and methods that evaluate multiple topologies and use an optimality criterion to pick the best tree.

Optimality criteria-based methods are far superior to methods like NJ as they evaluate all possible tree topologies. See Hillis et al. (1996) and Lemey et al. (2009) for more details regarding these tree-building methods.

References

Fitch WM, Margoliash E (1967) Construction of phylogenetic trees. Science 155:279–284

Lemey P, Salemi M, Vandamma A-M (2009) The phylogenetic handbook: a practical approach to phylogenetics analysis and hypothesis testing, 2nd edn. Cambridge University Press

Hillis DM, Moritz C, Mable BK (1996) Molecular systematics. Oxford University Press

Nucleotide Substitution Models and Estimating Pairwise Genetic Distance

4

Abstract

This chapter deals with the estimation of pairwise genetic distances using various nucleotide substitution models. Simple proportional differences (*p*-distance) are contrasted with corrected measures that account for multiple hits, saturation and varying substitution rates. Examples demonstrate how *p*-distance underestimates divergence in distantly related taxa. The chapter then explores different substitution models, their assumptions and applications, stressing the importance of choosing appropriate models for robust phylogenetic inference. Model selection is positioned as an essential prerequisite for likelihood and Bayesian approaches discussed later, highlighting the interplay between evolutionary processes and computational methods.

Keywords

Nucleotide substitution · Saturation · Multiple hits · Model selection · Corrected distance

Pairwise Genetic Distance

Thus far, we have been calculating pairwise genetic distances (*d*) by simply counting the number of nucleotides that are different between the two sequences. However, a more useful and informative measure of genetic distance is *p*-distance or proportional difference, which is calculated by dividing the total number of differences by the sequence length. Here genetic distance is normalised by sequence length. Thus, *p*-distance gives us a measure of the number of substitutions per site and can be compared across genes for a species pair. Additionally, it can be used to compare rates of evolution (substitution) across genes, i.e. genes that are evolving rapidly will show higher *p*-distances than those that are evolving slowly for a given taxonomic group.

In the example shown in Fig. 4.1, a 20 bp region has been sequenced from four species (S1–S4). The nucleotides in bold represent the sites that are different in

K. P. Karanth, *Molecular Phylogenetics*, https://doi.org/10.1007/978-981-95-9147-3_4

a)

S1 AACTGCATGGTAACAGGTTC 20
S2 AA**G**TGCATGGTAACAG**A**TTC 20
S3 AA**G**TG**G**ATGGTAA**T**AG**A**TTC 20
S4 AA**G**TG**GAC**GGT**T**A**T**AG**A**TTC 20

b)

	S1	S2	S3	S4
S1	0	0.1	0.2	0.3
S2	2	0	0.1	0.2
S3	4	2	0	0.1
S4	6	4	2	0

$$p \text{ distance}= \frac{\text{Number of substitutions between a pair of sequences } (d)}{\text{total length of the sequence}}$$

Fig. 4.1 Calculating *p*-distances. (**a**) Hypothetical sequence alignment of 20 nucleotides from species S1–S4. Nucleotides in bold represent sites that are different in S2–S4 when compared with S1. (**b**) Pairwise nucleotide difference (below diagonal) and *p*-distance (above diagonal) between species

S1 AACTGCATGGTAACAGGTTC 20
S2 ???[GCATGGTAA]CAG**A**TTC 16
S3 AA**G**[TG**G**ATGGTAA]??????? 13
S4 AA**G**TG**G**A?????A**T**AG**A**TTC 15

	S1	S2	S3	S4
S1	0	0.06	0.15	0.26
S2	1	0	0.11	0.18
S3	2	1	0	0
S4	4	2	0	0

Fig. 4.2 Calculating *p*-distances for partial sequences. (**a**) Hypothetical sequence alignment of 20 nucleotides from species S1 to S4. Missing data are depicted as "?". The overlapping region between S2 and S3 is boxed. Nucleotides in bold represent sites that are different in S2 to S4 when compared with S1. (**b**) Pairwise nucleotide difference (below diagonal) and *p*-distance (above diagonal) between species

species S2–S4 when compared with S1. To the right of the alignment is a matrix that shows pairwise genetic difference (below the diagonal) and *p*-distance (above the diagonal). For example, the number of differences between S1 and S2 is two nucleotide substitutions, which translates into 0.1 (2/20) *p*-distance.

Often, we have partial sequences for some species; to control for sequence length, the length of the overlap between sequences is used to calculate the *p*-distance (Fig. 4.2). For example, there is one difference between S2 and S3 in the overlapping region. The corresponding *p*-distance would be 0.11, i.e. one divided by nine (length of the overlap).

Multiple Hits and the Problem of Saturation

p-distance is a good measure of genetic distance for closely related and recently diverged species. However, for distantly related species, *p*-distance tends to underestimate genetic distance. This is because, with time, multiple changes might occur at the same site. Each time a substitution occurs at a site, it masks changes that occurred earlier at that site.

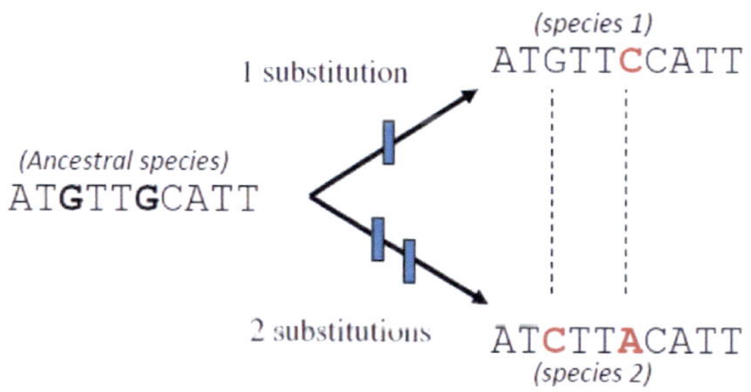

	Observed	Actual
Difference	2	3
p distance	0.2	0.3

Fig. 4.3 Actual and observed *p*-distances. Nucleotides in bold have undergone substitutions

Take the example illustrated in Fig. 4.3; here, an ancestral species has diverged into two daughter species. With time, a given region of the genome in these two species will begin to accumulate changes (begin diverging). When we compare their sequences today, a total of two differences are observed (shown as a dashed line), the first at site 3 and the second at site 6. Thus, the observed *p*-distance is 0.2 for this pairwise comparison. Nevertheless, it must be noted that these two changes have very different histories. Let us assume that we knew the sequence for this region in the ancestral species. This ancestral sequence can now be compared with extant (current) sequences to trace these changes or nucleotide substitutions (or simply substitutions). In the case of site 3, the observed difference between species 1 and 2 is due to a substitution along the lineage leading to species 2, and in species 1, no change has occurred. In contrast, in the case of site 6, one substitution each has occurred in lineages leading to both species 1 and 2. These include a substitution from G to C on the lineage leading to species 1 and a G to A on the lineage leading to species 2. Therefore, at site 6, two substitutions have occurred, but we have counted only one change. Such multiple substitutions at a site are called 'multiple hits'. Multiple hits at a site contribute towards underestimation of the *p*-distance because we cannot detect them when we compare extant sequences. Thus, the actual number of differences between these extant sequences should be three, whereas we have observed only two.

In Fig. 4.4, other examples of different kinds of substitutions at site 3 are illustrated. The first example (Fig. 4.4a) is the same as the one discussed in Fig. 4.3. Here, the actual $d =$ observed d; therefore, we do not underestimate d. However, in all other cases, we underestimate d due to multiple hits. It must be noted that these multiple hits have either occurred along one lineage (Fig. 4.4b, c) or have occurred along both the lineages (Fig. 4.4d, e). Additionally, some of these substitutions contributed towards genetic distance (Fig. 4.4a, b, e), while others make the daughter sequences similar at this site (Fig. 4.4c, d). An important component of genetic divergence is time; with time, sequences diverge from each other (accumulate changes). Additionally, with time, a given site is more likely to accumulate multiple hits.

If we plot *p*-distance for a sequence pair against time since divergence, we expect these sequences to accumulate changes and diverge from each other till *p*-distance is 1 or 100% difference (dashed line in Fig. 4.5). However, in nature, this never happens because after a prolonged period of time, the same site begins to accumulate

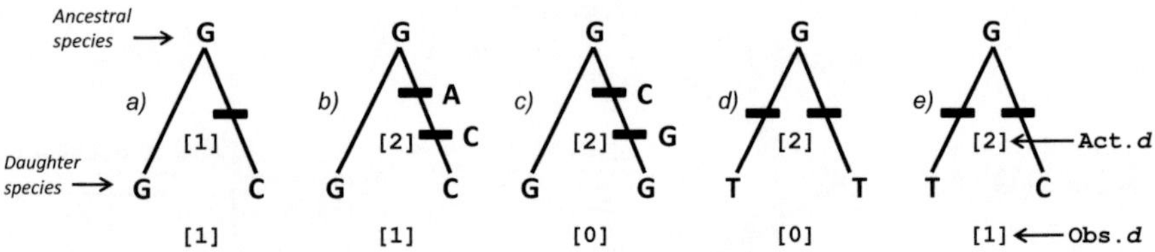

Fig. 4.4 Actual and observed genetic distance (numbers in square brackets) at a site due to single or multiple substitutions

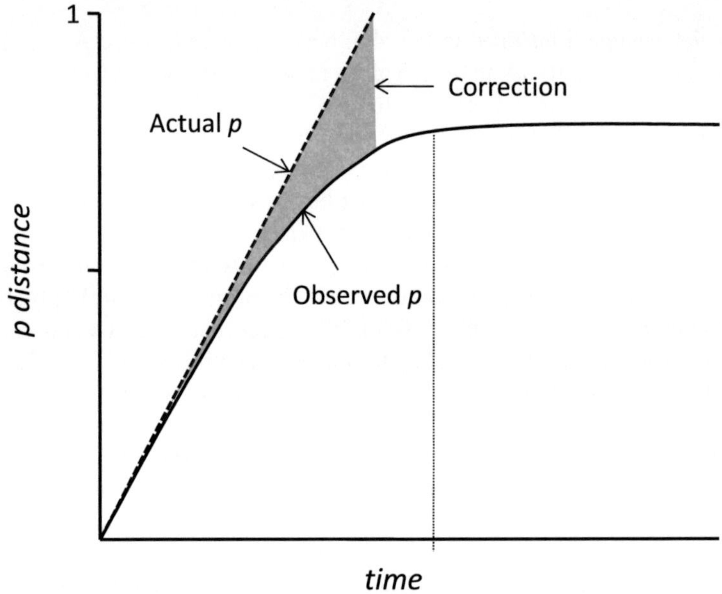

Fig. 4.5 *p*-distance vs. time

multiple substitutions (multiple hits). These multiple substitutions will mask earlier changes, and some changes will make the daughter sequences similar; thus, the observed distance will always be below the actual distance. Typically, the observed distance will increase rapidly initially but after a long time hit a plateau, i.e. it will attain saturation. At saturation, p will remain at the same value even though the sequences are still accumulating changes with time. This is because when p attains saturation, substitutions that contribute to divergence (Fig. 4.4a, b, e) will equal substitutions that make them similar (Fig. 4.4c, d). Thus, when p-distance is large it gives an underestimate of actual divergence.

Models of Sequence Evolution

Clearly, to get a better estimate of genetic distance (d), we need to correct for multiple hits; thus,

Actual d = Observed d + correction or p + correction.

Fortunately, the actual d can be estimated by incorporating correction to the observed d. The correction is based on one of various hypotheses about the nature of the nucleotide substitution process (models). These substitution/evolutionary models are a set of assumptions about the process of nucleotide substitutions.

The simplest model of the nucleotide substitution is the Jukes-Cantor model (Jukes and Cantor 1969). This model corrects for multiple hits but assumes that all substitutions are equally likely (discussed in more detail below) and that the base frequencies in a sequence are equal, i.e. proportions of A = T = C = G = 0.25. Additionally, the model assumes that the base frequencies are in equilibrium i.e. do not change over time. The estimate for d (nucleotide substitution/site) after incorporating the correction for multiple hits is given below.

$$d = -\tfrac{3}{4}\ln\left(1-\tfrac{4}{3}p\right)$$

Here, p is the observed p-distance. For information on how to derive this equation, *see* Chap. 3 of Nei and Kumar (2000).

As mentioned above, in the Jukes-Cantor model, all substitutions are equally likely. Figure 4.6 illustrates the various kinds of substitutions. These substitutions can be broadly categorised into transitions (α) and transversions (β). Transitions represent substitutions among purines or among pyrimidines, whereas transversions are substitutions across purines and pyrimidines. Thus, the p-distance consists of transitions and transversions or $p = \alpha + \beta$. The Jukes-Cantor model assumes that for a given dataset, $\alpha = \beta$, but in nature, transitions always outnumber transversions ($\alpha > \beta$). For example, in the case of vertebrate, mitochondrial gene transitions occur around eight times more often than transversions ($\alpha = 8\beta$). Therefore, multiple hits are a bigger issue in the case of transitions as they occur more often. The 'Kimura 2 parameter model' corrects for this transition bias by considering transitional and transversional substitutions separately. Here, two parameters α and β have to be calculated separately.

Thus, the new estimate for d is

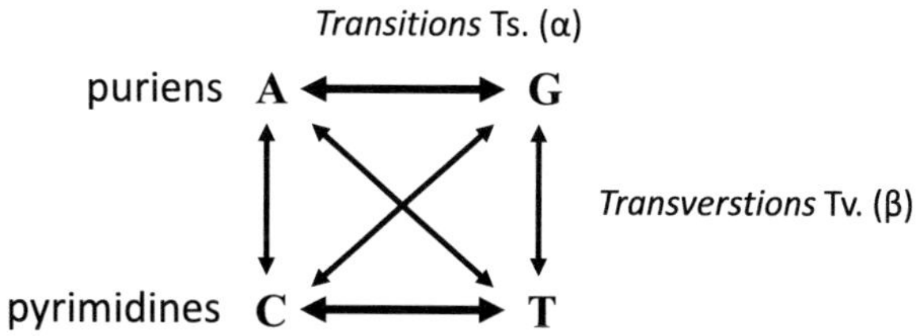

Fig. 4.6 Two kinds of substitutions: Transitions and transversions are represented by thick and thin arrowed lines, respectively

Fig. 4.7 Base composition in 1140 bp mitochondrial cytochrome *b* gene sequences from various primate species

	% base composition			
	T	C	A	G
Gibbon	24.3	33.8	28.3	13.6
Human	25.2	34.3	28.7	11.9
Macaque	25.2	34.0	29.1	11.7
Bonobo	25.2	34.1	29.2	11.5
Gorilla	24.8	34.6	28.9	11.7
Chimp	24.9	34.4	29.0	11.7
Average	24.9	34.2	28.9	12.0
Expected	25	25	25	25

$$d = -\frac{1}{2}\text{in}\left[\left(1-2\alpha-\beta\right)\sqrt{1-2\beta}\right]$$

For information on how to derive this equation, *see* Chap. 3 of Nei and Kumar (2000). Nevertheless, the Kimura 2 parameter model (K2P) assumes that the base frequencies in a sequence are in equilibrium. This assumption is also very often violated. For example, in the case of mitochondrial cytochrome *b* gene of humans (Fig. 4.7), if we count the number of A, T, G, C, then the equal base frequency assumption would predict equal numbers of these bases A = T = G = C, and their proportions would be 0.25. However, from Fig. 4.7, it is apparent that there is significant deviation from the expected 25% for all bases except T. The same is the case with other primate species.

To correct for unequal base composition, the Felsenstein (1981) model (also called F81) is used, but this model does not correct for transitional bias. Perhaps one of the most useful models is HKY 85 (Hasegawa et al. 1985); this model corrects for both transitional bias and unequal base composition. This model is commonly used for mitochondrial genes, where these corrections need to be incorporated. Table 4.1 lists the various models and their assumptions; this list is not exhaustive. As we go down the list, we keep adding more parameters to our estimate of *d*, and the models get more complex.

Among Site Rate Variation

All the above models make another major assumption, that is, rates of nucleotide substitutions across sites are the same or rates are homogeneous across site. This assumption is also often violated, particularly in the case of coding sequences. Take the example illustrated in Fig. 4.8. Here, a particular gene has been sequenced from four different species. The protein sequence coded by this gene is identical across species (conserved) and is given at the bottom of the alignment. However, the DNA sequences of this gene from the four species show considerable variation. This is

Table 4.1 Various substitution models and their assumptions. All base frequencies are at equilibrium

Models	Assumptions
Jukes and Cantor (1969) (JC)	All substitution equally likely, equal base frequencies
Kimura (1980) (K2P)	Substitutions are categorised into transitions and transversions, equal base frequencies
Felsenstein (1981) (F81)	All substitution equally likely, unequal base frequencies
Hasegawa et al. (1985) (HKY 85)	Substitutions are categorised into transitions and transversions, unequal base frequencies
General time reversible (GTR)	All six substitution classes have different rates, unequal base frequencies

		*	*		*	*	*		
Species 1	ATG	GGC	ACT	TAT	CCA	ATT	TTA	TAA	24
Species 2	ATG	GGT	ACT	TAT	CCC	ATC	TTA	TAA	24
Species 3	ATG	GGT	ACA	TAT	CCT	ATC	CTA	TAA	24
Species 4	ATG	GGC	ACG	TAT	CCT	ATT	CTA	TAA	24
AA	Met	Gly	Thr	Tyr	Pro	Ile	Leu	Stop	7

Fig. 4.8 Hypothetical sequences of four species, with variable sites marked by *, AA is amino acid

because all of the substitutions in the DNA alignment are silent/synonymous changes, or changes that do not alter the protein. Synonymous substitutions are largely confined to the third position of the codon.

In light of this information, we can revisit the assumption of rate homogeneity across sites. If this assumption were true, then the number of substitutions in the first, second and third positions of the codon should be equal. As we have seen with the example illustrated above, this is not the case. Here, substitutions have occurred predominantly at the third position of the codon; thus the rate of nucleotide substitution is higher for the third position followed by first position. Clearly, there is variation in substitution rates across sites. Similarly, substitution rates can vary across regions in a gene, for example, between exons and introns. This is called among-site rate variation, and this rate variation also needs to be incorporated in our estimate of *d*. To correct for among-site rate variation, we use gamma (Γ) correction as defined by the shape parameter called alpha (α). When $\alpha > 1$, the gamma shape parameter attains a bell-shaped curve; this means the dataset has low among-site rate variation, or substitution rates are, by and large, homogeneous across sites. On the other hand, when $\alpha < 1$, the gamma shape parameter attains an 'L' shaped curve; this means that there is among-site rate variation, with few sites exhibiting high substitution rates and most site exhibiting very low rates (Fig. 4.9).

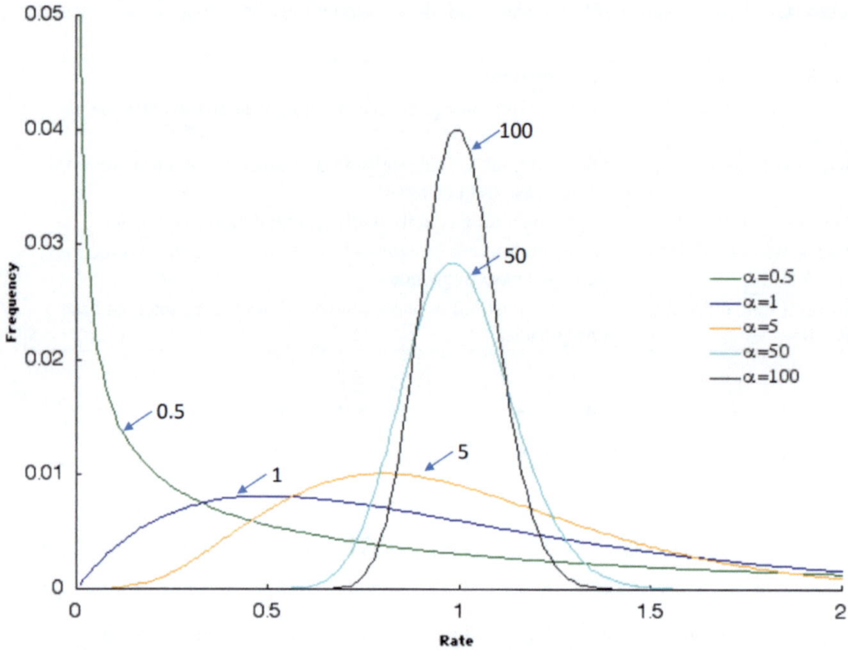

Fig. 4.9 Gamma shape parameter alpha (α). The numbers in the graph represent various alpha values

Other assumptions:

1. All nucleotide sites change independently of each other.
2. The substitution rate is constant over time and across lineages (molecular clock assumption).
3. Base composition across taxa is at equilibrium.

Nevertheless, it must be noted that the corrected *d* is an estimate of actual *d* and not necessarily equal to actual *d*. For a detailed treatment of this subject with a complete list of models and how they are related to each other, *see* the references listed at the end of this chapter.

Model Selection

The nature and rate of nucleotide substitutions in different segments of the genome vary depending upon selection acting on these regions. Thus, the issue of multiple hits and hidden changes vary across the genome. For example, for a neutrally evolving region that has no selective constraint, multiple hits at a site are more likely than a region coding for a protein (gene). Therefore, no single model will accurately capture the nature of these changes across a genome. Similarly, the nature and pattern of nucleotide substitutions will also vary across different genes. For a given gene alignment, one has to choose the model that best captures the nature of the nucleotide substitution process in that dataset. This is the first step before proceeding to build trees using the distance approach. Additionally, in the case of likelihood and Bayesian approaches, one has to specify an explicit model of sequence evolution (discussed in the next chapter). Thus, model selection is an important component of tree building. Given that there are many models, how do we choose the model that best fits our data? This aspect of tree building is dealt with in Chap. 9.

References

Felsenstein J (1981) Evolutionary trees from DNA sequences: a maximum likelihood approach. J Mol Evol 17(6):368–376

Hasegawa M, Kishino H, Yano T (1985) Dating of the human-ape splitting by a molecular clock of mitochondrial DNA. J Mol Evol 22(2):160–174

Jukes TH, Cantor CR (1969) Evolution of protein molecules. New York: Academic Press. pp 21–132

Kimura M (1980) A simple method for estimating evolutionary rates of base substitutions through comparative studies of nucleotide sequences. J Mol Evol 16(2):111–120

Nei M, Kumar S (2000) Molecular evolution and phylogenetics. Oxford University Press

Methods of Tree Building Using DNA Sequence Data II: Character-State Approach

5

Abstract

This chapter introduces character-state approaches, where sequence sites are treated as independent evolutionary characters. Maximum parsimony (MP) is explained as a method selecting the tree with the fewest evolutionary changes. Maximum likelihood (ML) is introduced as a method that uses a probabilistic framework and explicit substitution models to identify the most likely tree. Bayesian inference (BI), which uses a likelihood framework but differs from ML in that it incorporates prior probabilities in calculating posterior probabilities, is also discussed. The Markov chain Monte Carlo approach is explained with respect to estimating the posterior probability of a tree. Detailed worked examples comparing MP and ML are illustrated to highlight the difference between parsimony and likelihood-based methods.

Keywords

Parsimony · Likelihood · Bayesian inference · MCMC · Character state

Character-State Approach (Discrete Methods)

In the character-state approach, the value of the optimality criterion of a tree is inferred directly from sequence data, i.e. there is no intermediate step of calculating genetic distance, as in the case of distance-based methods. In this approach, each site in the sequence alignment is considered an independent character that can attain multiple states such as A, T, G and C; therefore, it is called a character-state approach.

Take the example of the sequence alignment in Fig. 5.1, when we compare any two taxa, such as chimp and gorilla; there are three differences between them; thus, the genetic distance is three. However, the nature of these substitutions is quite different. In the case of character number 4, the change is from A to C (or vice versa), but in the case of characters 8 and 9, it is from C/G to T (or vice versa). The character-state-based methods will evaluate each of these sites independently, but

K. P. Karanth, *Molecular Phylogenetics*, https://doi.org/10.1007/978-981-95-9147-3_5

	1	2	3	4	5	6	7	8	9
Human	G	T	C	A	C	A	T	C	T
Chimp	C	T	C	A	C	A	T	G	T
Gorilla	C	T	C	C	C	A	T	T	C
Orang	A	T	A	C	G	A	G	C	C

Fig. 5.1 An alignment of DNA sequence from human and great apes. Numbers indicate the location of each site/character in the alignment

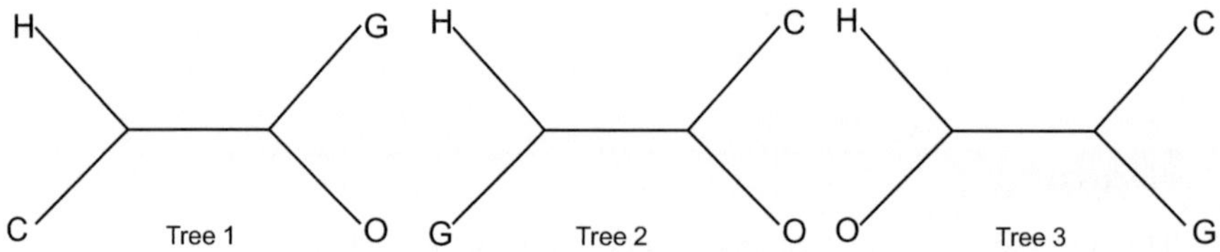

Fig. 5.2 Three possible unrooted trees in the case of four taxa (human: H, gorilla: G, orang: O, chimp: C)

the distance-based methods will club them into one, giving a genetic distance of three. If we now compare another species pair, such as human-chimp and chimp-gorilla, the characters that contribute to their genetic distances are again different. In the case of human-chimp, it is characters 1 and 8, whereas in the case of chimp-gorilla, it is characters 4, 8 and 9. Here, again, the distance-based approaches do not account for the difference in the characters that contribute towards genetic distances across species pairs.

Maximum Parsimony (MP)

A parsimony tree is one that needs the minimum number of nucleotide substitutions (changes) to explain the differences that are observed between the sequences. In this approach, we evaluate all possible trees for a given set of taxa and use the principle of parsimony to pick the tree requiring the least number of steps or changes to explain the character distribution. The tree requiring the least number of changes is called the optimal tree. Parsimony is the principle of endorsing the simplest explanation that resolves a case or a problem.

As mentioned earlier, there are multiple possible trees for a given number of taxa. From a previous exercise, we know that for four taxa, there are three possible unrooted trees. With respect to great ape-human relationships, these trees are illustrated in Fig. 5.2.

The question then is which among these three trees is the most parsimonious tree or optimal tree? To do this, we have to first calculate the minimum tree length, using the principle of parsimony, for each of the trees above, and pick the one with the shortest tree length. In this regard, MP is similar to other optimality criteria-based tree-building methods, such as ME and LS (*see* Chap. 3), in that we are attempting to optimise the value of the optimality criteria.

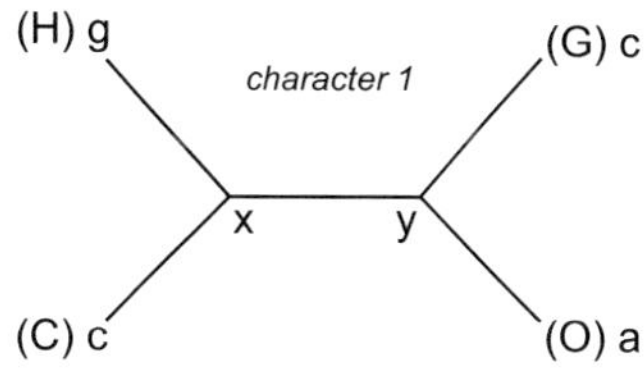

Fig. 5.3 Character 1 mapped onto the tips of tree1

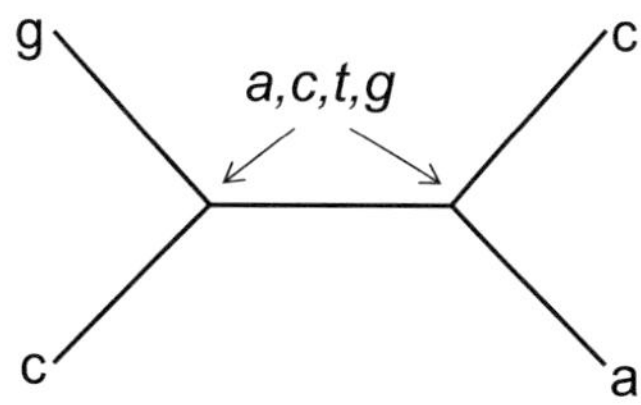

Fig. 5.4 Character states of internal node x and y

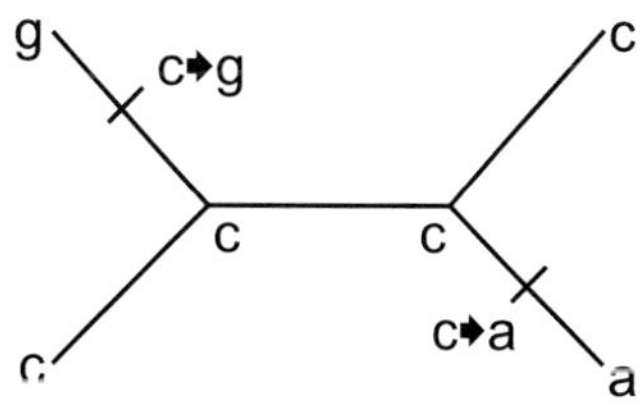

Fig. 5.5 Character states of internal nodes x and y are designated as 'c'. Cross bars represent a change from one state to another and arrows indicate the polarity of the change

Calculating Tree Length (TL)/Parsimony Score

To calculate the tree length, we have to reconstruct the evolution of each site/character of the alignment on a tree. Let us start with tree 1. First, we pick character 1 from the alignment, and the characters states are mapped onto the tips of tree 1.

In the case of character 1, chimp and gorilla have the character state 'c', whereas human and orange have 'g' and 'a', respectively (Fig. 5.3).

Next, we attempt to reconstruct the character states of the internodes (x and y) representing the common ancestor of HC and GO. The character state of x and y could be either of the four bases (a, t, g or c) as illustrated in Fig. 5.4.

Nevertheless, we are interested in the combination of character states for these internal nodes, such that the number of changes along the different lineages is minimised (principle of parsimony). It turns out that the optimal state is a 'c' for both nodes x and y. Take the case of node x; if the ancestral state at this node was a 'c', then we would have to invoke one change (substitution from 'c' to a 'g') on the lineage leading to human. Similarly, there would be one change (from 'c' to a 'a') on the lineage leading to orang. Thus, for character 1, two changes are invoked to explain the variation observed among species (Fig. 5.5).

If any other character states were assigned to the internodes x and y, then more than two changes would have to be invoked. For example, if character state 'g' and 'a' are assigned to nodes x and y, then we will have to invoke three changes to explain the observed differences (Fig. 5.6).

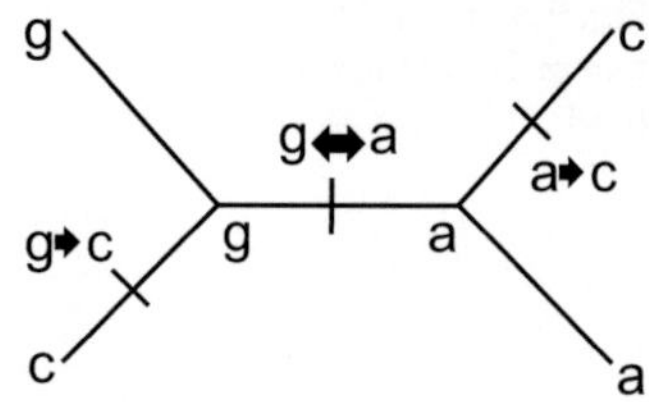

Fig. 5.6 Character state of internal node x is ‘g’ and y is ‘a’

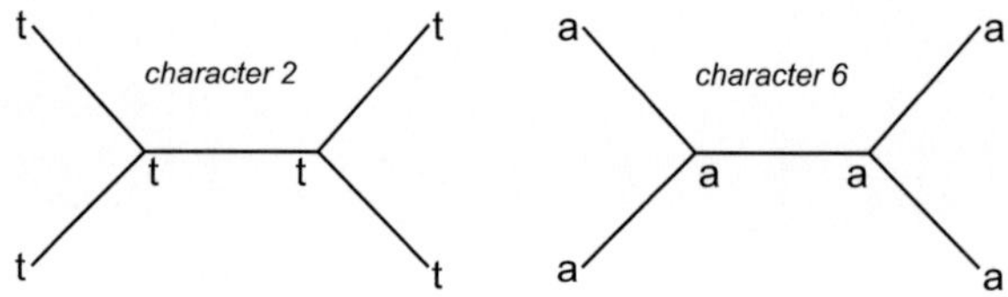

Fig. 5.7 Character states for internal nodes x and y for characters 2 and 6

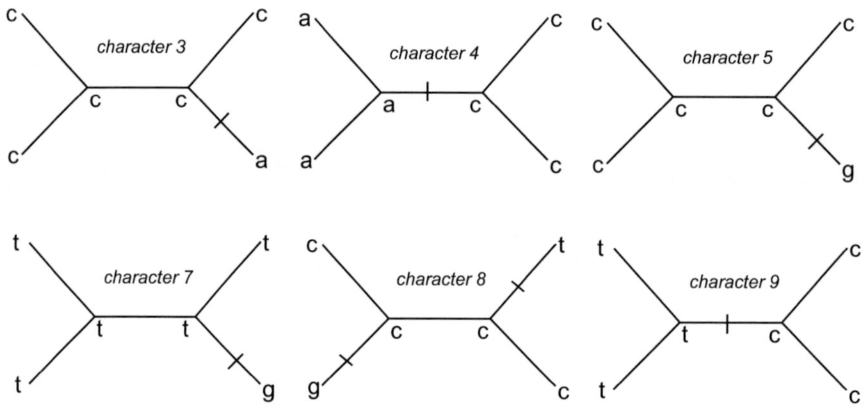

Fig. 5.8 The most optimal assignment of character states to internodes x and y for characters 3, 4, 5, 7, 8 and 9

Characters 2 and 6 have the same state (‘t’ and ‘a’) among all four taxa, and therefore no change (0) is invoked (Fig. 5.7).

For the remaining characters (3–5 and 7–9), we come up with the following ancestral character states (Fig. 5.8). Thus, characters 3–5, 7 and 9 contribute one change each to tree 1, whereas characters 1 and 8 contribute two changes each.

The sum of all these changes gives us the total tree length. Figure 5.9 summarises the contribution of each character towards the total tree length for tree 1. Thus, tree 1 needs nine substitutions to explain the variations seen among the sequences in the alignment.

We can also map these nine changes on to the five branches to give us the branch lengths for tree 1 (Fig. 5.10).

Thus, lineages leading to human, chimp and gorilla have one change each. There are four changes along the gorilla lineage and two between the nodes x and y. The

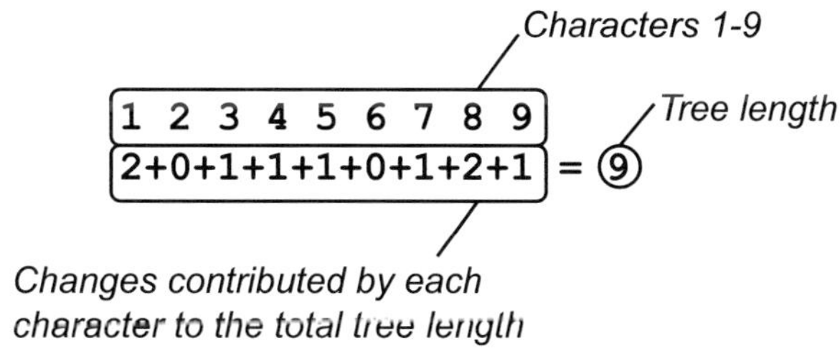

Fig. 5.9 Summary of the changes contributed by characters 1–9

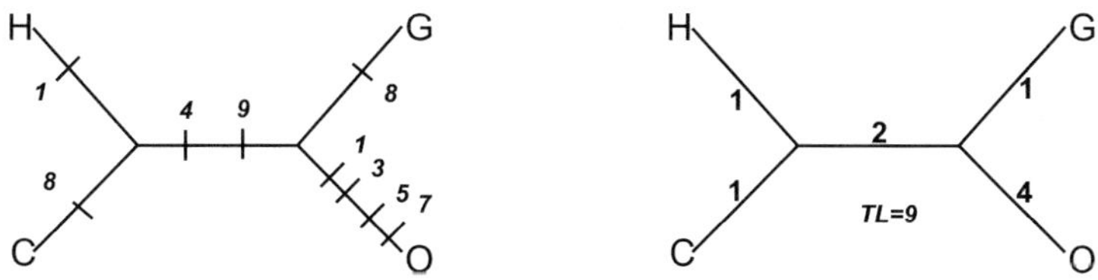

Fig. 5.10 Tree on the left shows the characters that have undergone changes along each branch; tree on the right has the number of changes mapped along each branch along with tree length (TL)

above-described steps have to be repeated on the other two trees to determine their tree lengths. Figure 5.11 compares the tree lengths of the three trees.

From this exercise, it is apparent that tree 1 is the optimal tree as it needs the least numbers of changes, (9) when compared to tree 2 and 3, which both need 11 changes. Thus, the most parsimonious tree among the three trees is tree 1 (Figs. 5.11 and 5.12).

The branch lengths of tree 1 can be scaled according to the number of changes to give the unrooted phylogram (Fig. 5.13). In this unrooted phylogram, the longest tree distance between any two taxa is seven, i.e. between human and orang or chimp and orang. The midpoint on this branch is at 3.5, along the orang lineage, which can be used for midpoint rooting. This tree can be rooted using midpoint rooting to give a rooted phylogram (Fig. 5.13).

The rooted cladogram depicting the same tree is shown Fig. 5.14.

See Stewart (1993) for a detailed treatment on the pros and cons of parsimony.

Maximum Likelihood

Likelihood is the probability of observing a certain outcome (observed data) given a particular hypothesis. Thus, Likelihood (L) = Probability (P) of data (D) given the hypothesis (H). Likelihood is also referred to as conditional probability because the probability of an event is contingent upon a given hypothesis. Likelihood (L) is described in the following equation:

$$L = \mathrm{P}\left(\mathrm{D}|\mathrm{H}\right)$$

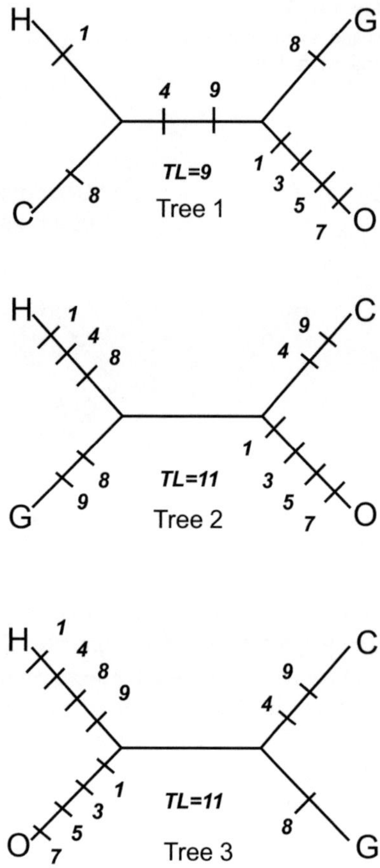

Fig. 5.11 Character changes mapped onto the three trees along with their tree lengths (TL)

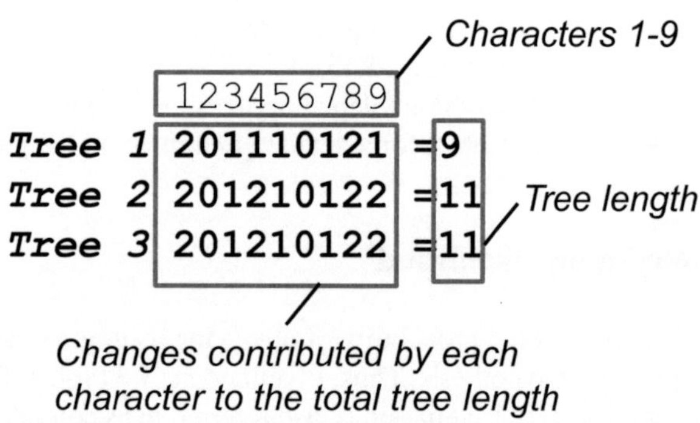

Fig. 5.12 Summary of the changes contributed by characters 1–9 to trees 1–3

Here, the vertical line (|) represents the condition aspect of likelihood. To better understand likelihood, let us take the coin tossing example. If we toss a coin four times and the outcome of this exercise is one heads (h) and three tails (t), the sequence of heads and tails is not important here. What is the likelihood of this

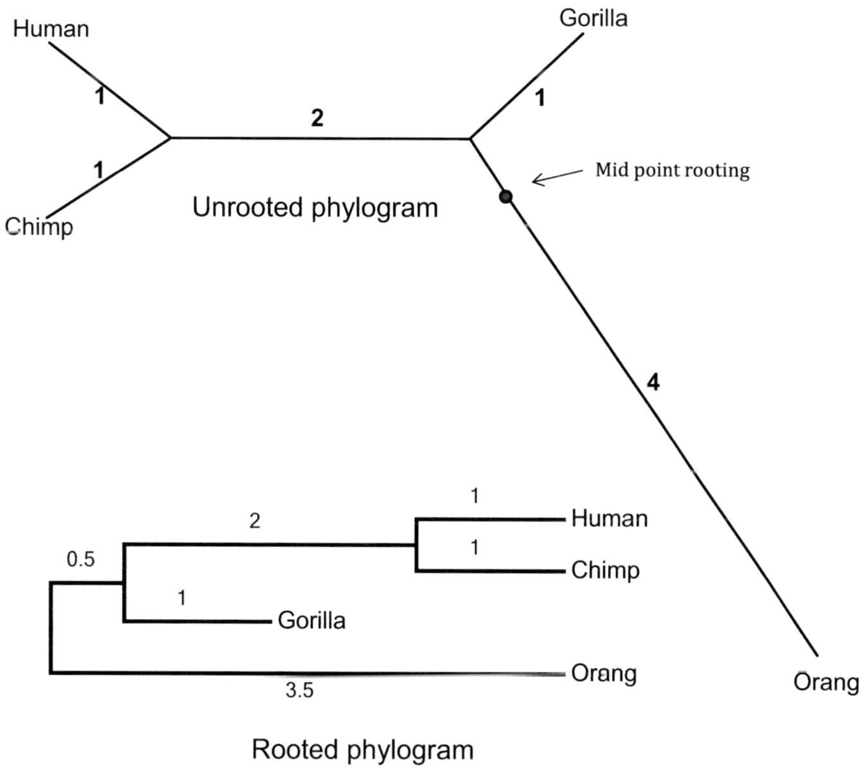

Fig. 5.13 Unrooted and rooted phylograms; numbers indicate changes along each branch

Fig. 5.14 Rooted cladogram

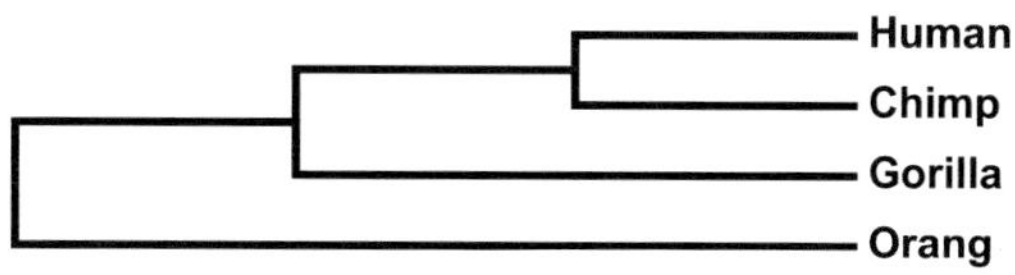

outcome? The answer depends on the hypothesis, i.e. whether or not the coin is fair. In this example, the data (D) are 1 h, 3 t and the hypothesis is probability (p) of getting heads/tails for any one flip of the coin. If the coin is fair, then the probability (p) of heads or tails is the same.

$$p(h) = p(t) = 0.5$$

However, p(t) = 1-p(h); thus, we need just one piece of information, either p(h) or p(t), for our hypothesis. We will first calculate the likelihood of the data (1 h, 3 t) given the hypothesis p(h) = 0.5 and we get the following:

$$P(1h, 3t) = p(h) \times [1 - p(h)] \times [1 - p(h)] \times [1 - p(h)]$$

$$(0.5)(1-0.5)(1-0.5)(1-0.5) = 0.5 \times 0.5 \times 0.5 \times 0.5 = 0.062$$

Thus, if the coin is fair, i.e. p(h) = 0.5, then the likelihood of the data (1h, 3t) is 0.062.

What about the likelihood of this data given the hypothesis p(h) = 0.25?

$$0.25 \times 0.75 \times 0.75 \times 0.75 = 0.10$$

The likelihood has gone up, and this is intuitive from the data. It appears that this coin is biased because we get tails in three out of four flips. We can now consider various hypotheses (H1–4), i.e. values of p(h). You can also come up with other hypotheses and calculate likelihoods of each.

$$\text{H1}: \text{p}(\text{h}) = 0.1 L = 0.072$$

$$\text{H2}: \text{p}(\text{h}) = 0.25 L = 0.10\ (\text{ML})$$

$$\text{H3}: \text{p}(\text{h}) = 0.5 L = 0.062$$

$$\text{H4}: \text{p}(\text{h}) = 0.75 L = 0.011$$

From this exercise, it is clear that the hypothesis that is most likely to generate the observed outcome is p(h) = 0.25; the maximum likelihood (ML) of this data given this hypothesis is 0.10. Thus, we have used the likelihood values of different hypotheses to pick the hypothesis that best fits the data. The maximum likelihood approach, therefore, attempts to maximise the likelihood of the observed data by evaluating various hypotheses and picking the hypothesis that is most likely to generate the observed data.

Maximum Likelihood Approach in Tree Building

The likelihood of a tree is the probability of generating the alignment given the tree and model of nucleotide substitution/sequence evolution. Thus, the sequence alignment is the data (D), and the given tree along with a model of nucleotide substitution is the hypothesis.

So instead of $L = \text{P}(\text{D} \mid \text{H})$

We have $L = \text{P}(\text{S} \mid \text{T} + \text{m})$

where S = sequence alignment, T = tree topology and m = model of sequence evolution.

The first step in determining the likelihood of a tree is model selection (Chap. 4). The same model is then used to determine the likelihoods of various trees for a given dataset. In the case of four-taxa alignment, we have three possible trees or hypotheses under the same model of nucleotide substitution.

$$H1 : T1 + m$$

$$H2 : T2 + m$$

$$H2 : T3 + m$$

Thus,

$$L_1 = P(S|T1 + m), L_2 = P(S|T2 + m), L_3 = P(S|T3 + m)$$

The tree topology is the only variable that changes; both model (m) and data (S) remain constant. We pick the tree that gives the highest likelihood score, and this is our ML tree. This will be the tree that is most likely to generate the observed variations among sequences in the alignment, given a particular model of sequence evolution.

Calculating the Likelihood of a Tree

Just like in parsimony, in the case of likelihood too, we start with calculating the likelihood of a tree for each site (character). For character 1, the following alignment has different states in the four taxa (Fig. 5.15).

These states are mapped onto the terminal nodes of tree 1 (from Fig. 5.2), with x and y representing the states for the internal nodes (Fig. 5.16) as we have done earlier with parsimony.

It is convenient to root this tree, and for our purpose, we place the root on the branch xy to give the rooted tree illustrated in Fig. 5.17.

In this tree, z represents the character state of the root and V1–V6 are branch lengths. The branch length V is the product of mutation rate (μ) and time (t).

Thus, $V = \mu t$.

The likelihood of a particular site is the summation of the probabilities of every possible reconstruction of ancestral states, given some model of base substitution and a tree.

Fig. 5.15 An alignment of DNA sequence from human and great apes. Numbers indicate the location of each site/character in the alignment

	1	2	3	4	5	6	7	8	9
Human	G	T	C	A	C	A	T	C	T
Chimp	C	T	C	A	C	A	T	G	T
Gorilla	C	T	C	C	C	A	T	T	C
Orang	A	T	A	C	G	A	G	C	C

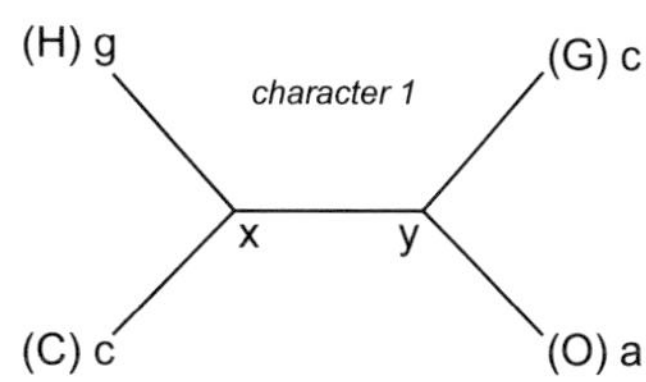

Fig. 5.16 Character 1 mapped onto the tips of tree1

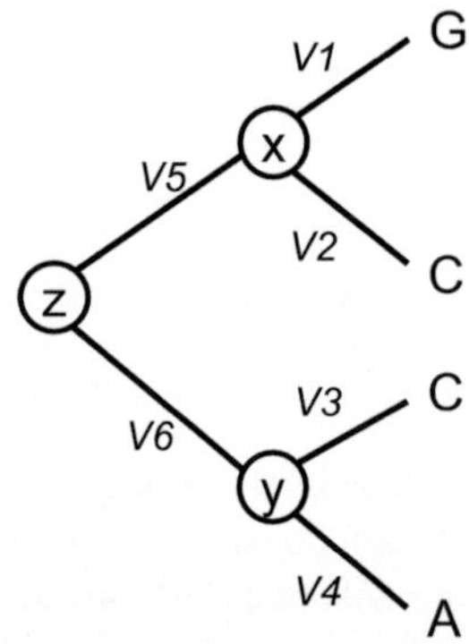

Fig. 5.17 Rooted tree 1 with taxa at the tips being replaced by various states of character 1. V1–V6 are branch lengths. X, Y and Z represent internal nodes

The internal nodes x, y and z could have been any one of the four bases (A, T, G or C). Given this background, the likelihood of tree 1 generating the observed distribution of bases for character 1 is

$$L = fz \cdot P_{ZX}(V5) \cdot P_{XG}(V1) \cdot P_{XC}(V2) \cdot P_{ZY}(V6) \cdot P_{YC}(V3) \cdot P_{YA}(V4)$$

We start with the root z, where fz is the prior probability of base z. From the root, we go down the different branches. $P_{ZX}(V5)$ is the probability of substitution from state z to x, given the branch length V5, as we travel from node z to x. Similarly, on reaching x, we travel down the two branches leading to the tips $P_{XG}(V1)$ and $P_{XC}(V2)$. This process is repeated along the branch leading to y. The probability of substitution from state i to state j, given branch length V, can be written as $P_{ij}(V)$ (where i = A/T/G/C and j = A/T/G/C). In the case of Jukes-Cantor, the simplest model of nucleotide substitution is

$$P_{ij}(V) = (1/4)\left(1 - e^{-v}\right)$$

The probability of no change, given branch length *V*, can be written as *Pii*(*V*).

$$P_{ii}(V) = (1/4) + (3/4)\left(e^{-v}\right)$$

However, calculating the likelihood of this tree is computationally intensive, given we do not know x, y and z. These internal nodes, as mentioned earlier, could have been A, T, G or C (Fig. 5.18).

Thus, the likelihood of this tree is summation over all possible states (A, T, G, C) at nodes x, y and z.

$$\sum_{Z}\sum_{X}\sum_{Y}\left(fz \cdot P_{ZX}(V5) \cdot P_{XG}(V1) \cdot P_{XC}(V2) \cdot P_{ZY}(V6) \cdot P_{YC}(V3) \cdot P_{YA}(V4)\right)$$

where X, Y, Z = A, T, G and C.

Felsenstein (2003) introduced a shortcut, called tree pruning, to simplify the computation of likelihood of a tree. In the tree pruning method, we start with nodes that have tips as their immediate descendants and work our way to the root. To better understand tree pruning, let us start with the sub-tree consisting of node x and the tips, human and chimp: these tips have bases G and C, respectively (Fig. 5.19).

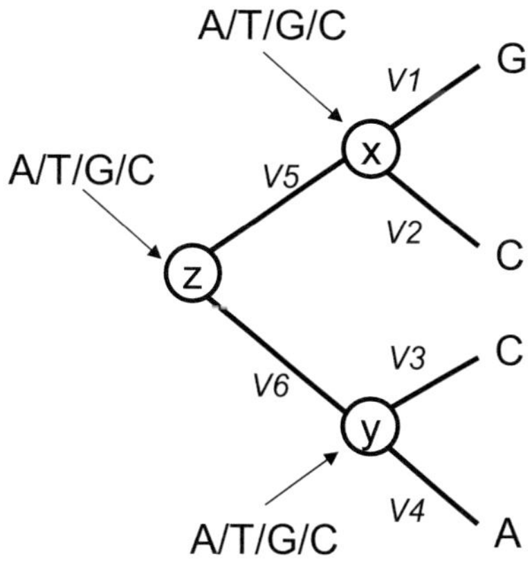

Fig. 5.18 Rooted tree 1 wherein the internal nodes X, Y and Z could be A or T or G or C

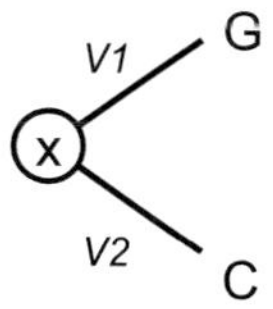

Fig. 5.19 Subtree consisting of node x and daughter lineages

This part of the tree is represented by the section of our likelihood equation (in bold)

$$\sum_{Z}\sum_{X}\sum_{Y}\left(fz \cdot P_{ZX}\left(V5\right) \cdot \mathbf{P_{XG}}\left(\mathbf{V1}\right) \cdot \mathbf{P_{XC}}\left(\mathbf{V2}\right) \cdot P_{ZY}\left(V6\right) \cdot P_{YC}\left(V3\right) \cdot P_{YA}\left(V4\right)\right)$$

$[P_{XG}(V1)]$ $[P_{XC}(V2)]$, where x is A, T, G, C.

Therefore, we have four possibilities:

If X = A, then $[P_{AG}(V1)]$ $[P_{AC}(V2)]$
If X = G, then $[P_{GG}(V1)]$ $[P_{GC}(V2)]$
If X = T, then $[P_{TG}(V1)]$ $[P_{TC}(V2)]$
If X = C, then $[P_{CG}(V1)]$ $[P_{CC}(V2)]$

Given $P_{ij}(V) = (1/4)(1 - e^{-v})$ and $P_{ii}(V) = (1/4) + (3/4)(e^{-v})$ for a branch length V, we calculate P_{ij} and P_{ii} for various values of V, for example, for $V = 0.4$, P_{ii}=0.75 and P_{ij}=0.08. These values can be used to generate the substitution probability matrix, given $V = 0.4$ (Table 5.1).

Using this matrix and assuming that V1 = V2 = 0.4, we get the following values:

$$P_{ij}\left(V1\right) = P_{AG}\left(0.4\right) = P_{TG}\left(0.4\right) = P_{CG}\left(0.4\right) = 0.08$$

$$P_{ij}\left(V2\right) = P_{AC}\left(0.4\right) = P_{GC}\left(0.4\right) = P_{TC}\left(0.4\right) = 0.08$$

$$P_{ii}\left(V1\right) = P_{GG}\left(0.4\right) = P_{ii}\left(V2\right) = P_{CC}\left(0.4\right) = 0.75$$

Table 5.1 Substitution probability matrix

	A	T	G	C
A	0.75	0.08	0.08	0.08
T	0.08	0.75	0.08	0.08
G	0.08	0.08	0.75	0.08
C	0.08	0.08	0.08	0.75

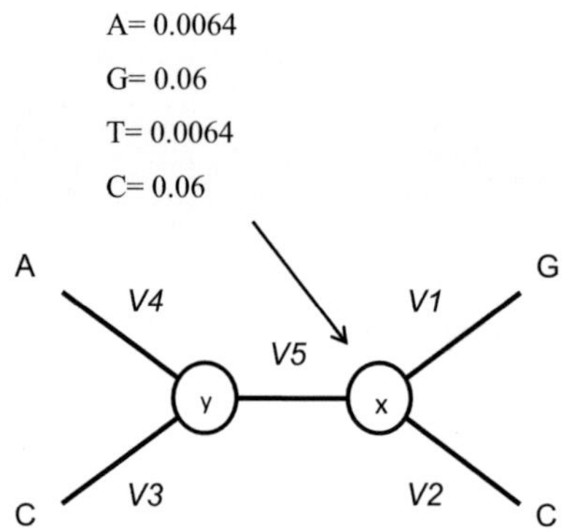

Fig. 5.20 Probability of node X being A/G/T/C

Thus, for the internal x, we get the following values:

$$[P_{AG}(V1)][P_{AC}(V2)] = 0.08 \cdot 0.08 = 0.0064$$

$$[P_{GG}(V1)][P_{GC}(V2)] = 0.75 \cdot 0.08 = 0.06$$

$$[P_{TG}(V1)][P_{TC}(V2)] = 0.08 \cdot 0.08 = 0.0064$$

$$[P_{CG}(V1)][P_{CC}(V2)] = 0.08 \cdot 0.75 = 0.06$$

Thus, given that the tips following node X have G and C, the probability that X is A/G/T/C is 0.0064, 0.06, 0.0064 and 0.06, respectively (Fig. 5.20).

Similarly, we calculate for node Y (Fig. 5.21).

We now have the probability of various bases for the internal nodes x and y. Next, we need to calculate the probability of change from a particular base to another base (Pij), as we move from one internal node to another internal node. We could either move from x to y or vice versa, depending on which internal node is taken as the root. The model used here is time reversible, so $P_{ij} = P_{ji}$, and therefore it does not matter which node is taken as the root. For our purpose, we take node y as the root, giving us the rooted tree depicted in Fig. 5.22.

Our tree is now reduced to only two nodes, x and y. The likelihood of this tree is $P_{XY}(V5)$, which is the probability of changes from x to y summed over all bases, given branch length V5. Here, again, we assume V5 = 0.4.

For example, if Y = A and X = A, then:

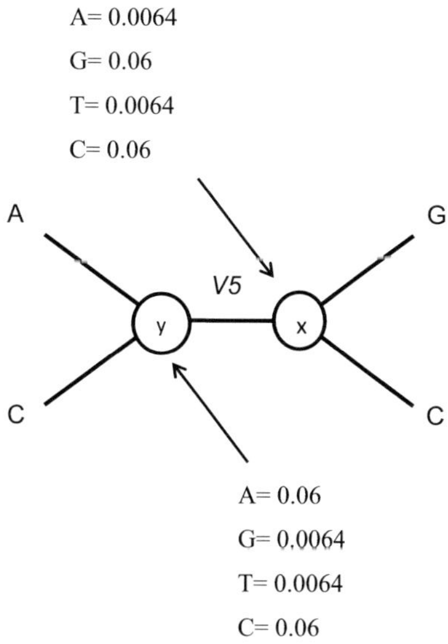

Fig. 5.21 Probability of node Y being A/G/T/C

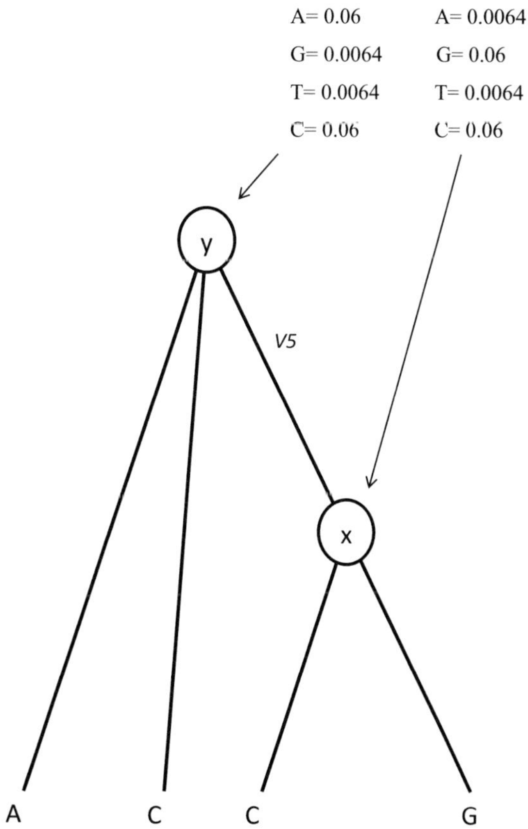

Fig. 5.22 Rooted tree with node Y serving as the root

P_{AA} = (prior probability of A at node X) x (probability of no change) x (prior probability of A at node Y)

Thus,

$$\text{Y} = \text{A X} = \text{A}\, 0.0600 \times 0.75 \times 0.0064 = 0.000288$$

Similarly,

$$\text{If Y} = \text{A X} = \text{G}\, 0.0600 \times 0.08 \times 0.0600 = 0.000288$$
$$\text{If Y} = \text{A X} = \text{T}\, 0.0600 \times 0.08 \times 0.0064 = 0.000030$$
$$\text{If Y} = \text{A X} = \text{C}\, 0.0600 \times 0.08 \times 0.0600 = 0.000288$$
$$\text{Total } 0.000894$$

The total gives us the probability that the root node (Y) is an A, given X is A/T/G/C.

Similarly, the probability that the root node (Y) is a G = 0.000324,

$$\text{T} = 0.000093 \text{ and C} = 0.003048$$

The likelihood of the tree, then, is the summation of probabilities of all four bases at the root node.

A	0.000894
G	0.000324
T	0.000093
C	0.003048
Total	0.004359

In our example, we have used the Jukes-Cantor model, which assumes the equilibrium base frequencies to be 0.25. Therefore, this total has to be multiplied by 0.25, which is the prior probability (fz from the likelihood equation) of a particular base at the root node.

Thus, the likelihood that this tree will generate the observed data for site 1 (A, C, C, G at the tips of the tree) given the Jukes-Cantor model is $0.004359 \times 0.25 = 0.00108$.

$$L = 0.00108$$

Similarly, we calculate the likelihood for all the sites:

$$\text{Site 1)}\ 0.004359 \times 0.25 = 0.00108$$
$$\text{Site 2)}\ 0.239140 \times 0.25 = 0.0597$$
$$\text{Site 3)}\ 0.029098 \times 0.25 = 0.00727$$
$$\text{Site 4)}\ 0.031943 \times 0.25 = 0.00798$$
$$\text{Site 5)}\ 0.029098 \times 0.25 = 0.00727$$
$$\text{Site 6)}\ 0.239140 \times 0.25 = 0.0597$$
$$\text{Site 7)}\ 0.029098 \times 0.25 = 0.00727$$
$$\text{Site 8)}\ 0.004359 \times 0.25 = 0.00108$$
$$\text{Site 9)}\ 0.031943 \times 0.25 = 0.00798$$

Then, the likelihood of a tree, given an alignment, is the product of the likelihoods of each site.

$$L = L(1) \times L(2) ..... \times L(9)$$

Thus, the likelihood of this tree, given the complete alignment is

$$L = 0.00108 \times 0.0597 \times 0.00727 \times 0.00798 \times 0.00727 \times 0.0597 \times 0.00727 \times 0.00108 \times 0.00798$$

Since the individual likelihoods are extremely small numbers, their product is difficult to calculate. Thus, it is convenient to convert to logarithm and sum the log likelihoods at each site to report the likelihood of the entire tree as the log likelihood.

$$\ln L = \ln L(1) + \ln L(2) ..... + \ln L(9)$$

Thus, the log likelihoods of each site are

	L	ln L
1.	0.00108	−6.83
2.	0.0597	−2.81
3.	0.00727	−4.92
4.	0.00798	−4.83
5.	0.00727	−4.92
6.	0.0597	−2.81
7.	0.00727	−4.92
8.	0.00108	−6.83
9.	0.00798	−4.83
Total		−43.70

The log likelihood (ln L) of tree 1 is the summation of the log likelihoods of each site, and thus we get −43.70. We can now calculate the likelihood of the other two trees to determine if we can get a better score. However, it must be noted that we have taken the branch lengths to be constant across sites and lineages, i.e. V1–V5 = 0.4. Therefore, for each site, we have assumed the branch lengths shown in Fig. 5.23.

This is an unreasonable assumption, given that the rate of evolution can vary across lineages for a site and across sites, in an alignment. For example, for invariant sites (sites 2 and 6), we get the highest likelihood value when V = 0. Thus, to get the best likelihood score for a site, we need to vary the branch lengths and pick the combination of branch lengths that gives us the highest value. These calculations are computationally intensive and difficult to do by hand. Here, I have used the programme *PAUP* (Swofford 2003) to calculate the likelihood of the three unrooted trees (Fig. 5.24).

The maximum likelihood tree, then, is tree 1, which has the highest likelihood score. The likelihood scores calculated by *PAUP* are better than the ones we got for tree 1, because the program has evaluated different branch lengths for each site. The

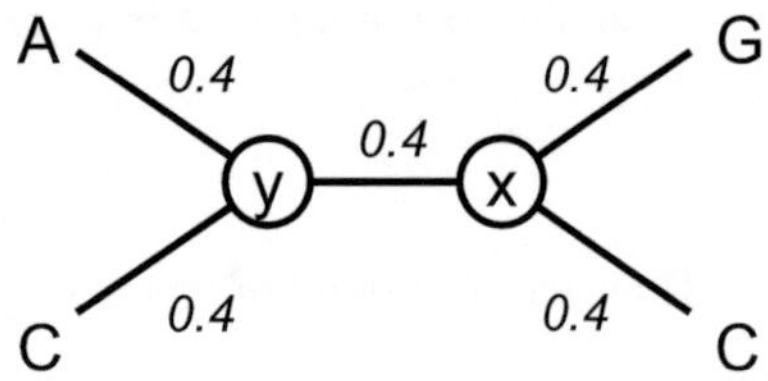

Fig. 5.23 Unrooted tree 1 with internal branch lengths (V = 0.4)

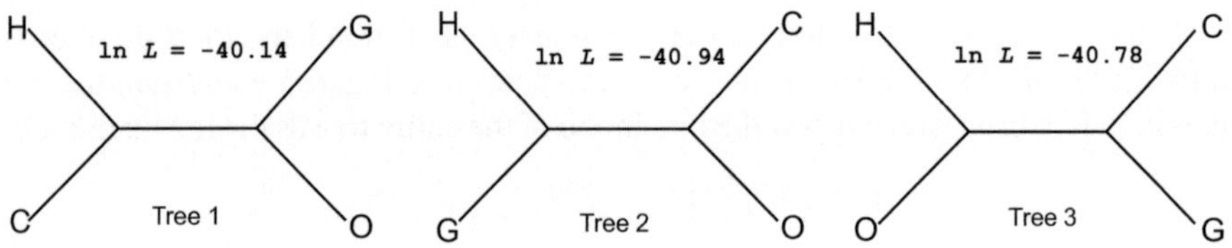

Fig. 5.24 Log likelihoods of the three trees

maximum likelihood tree (tree 1) can now be rooted to get the rooted cladogram. We can then ask the program to estimate the branch lengths and give us a phylogram.

Bayesian Approach

The Bayesian approach is closely allied to likelihood (Huelsenbeck et al. 2002; Holder and Lewis 2003), in that it uses a likelihood function. In likelihood, we are trying to determine the probability of the data given the hypothesis.

$$L = \mathrm{p}(\mathrm{D}|\mathrm{H})$$

Likelihood is difficult to interpret because it represents the probability of the data, given the hypothesis (Lewis 2001). Another interpretation of likelihood is that it is the probability that the data could have been generated by the hypothesis (tree topology gives a model of nucleotide substitution). However, it is more useful to determine the probability of the hypothesis because one can then evaluate support for different hypotheses. In the Bayesian approach, we determine the probability of the hypothesis given the data, i.e. p(H|D).

$$\mathrm{p}(\mathrm{H}|\mathrm{D}) = \frac{\mathrm{p}(\mathrm{H}) \times \mathrm{p}(\mathrm{D}|\mathrm{H})}{\mathrm{p}(\mathrm{D})}$$

Here, p(H|D) is the posterior probability of the hypothesis, p(H) = prior probability of the hypothesis, p(D|H) = Likelihood of the data and p(D) = probability of the data summation over all hypotheses.

To better understand the Bayesian approach, let's use an example. Let us consider two bowls, consisting of the same number of marbles, of two different colours in different proportions. Bowl A has 25 white and 25 black marbles, and bowl B has 10 white and 40 black marbles. These marbles are of the same size. If you were blindfolded and asked to pick a marble from one of the bowls that was randomly presented to you and the marble turns out to be a black marble, which bowl was it picked from? We have two hypotheses to consider: hypothesis 1 is that the marble was picked from bowl A, and hypothesis 2 is that it was picked from bowl B. Our intuition given the information is that you probably picked it from bowl B, given it has more black marbles. Thus, we place higher probability on hypothesis 2.

We can now apply Bayesian inference to determine the probability of hypothesis 2, given the data (black marble). The prior probability [p(H)] of hypotheses 1 is the same as hypothesis 2 and is equal to 0.5. This is because one of the bowls was randomly picked and presented to you, and each of the bowls has the same probability of being picked. The probability of the data, given hypothesis 1 is true [p(D|H1)], is 25/50 = 0.5; similarly, p(D|H2) is 40/50 = 0.8. The probability of the data p(D), then, is the sum of the probabilities of getting a black marble if either of the two hypotheses were correct. In other words, if we were to mix the contents of the two bowls, what is the probability of the observed data (getting a black marble)?

$$p(D) = p(H1) \times p(D|H2) + p(H2) \times p(D|H2)$$

The formula for Bayesian inference would be

$$p(H2|D) = \frac{p(H2) \times p(D|H2)}{p(H1) \times p(D|H1) + p(H2) \times p(D|H2)}$$

Thus,

$$p(D|H2) = 0.5 \times 0.8 / (0.5 \times 0.5 + 0.5 \times 0.8) = 0.4 / 0.65 = 0.61$$

Similarly, p(D|H1) = 0.38. Bayesian inference (BI) has confirmed our suspicion that the marble was most probably picked from bowl B. Thus, in BI, we use prior information regarding the probability of the hypothesis and modify our belief in the light of new data to come up with a posterior probability of the hypothesis. In our example, our prior belief based on available data was that a marble of any colour could be picked from either bowl A or B [p(H1) = p(H2) = 0.5]; however, on picking a black marble (new data), we have modified our belief (posterior probability) and placed higher probability on hypothesis 2. Thus, we went from p(H2) = 0.5 to p(H2|D) = 0.61. Even though both ML and BI calculate probabilities and use likelihood functions, the fundamental difference between them is the use of priors in the case of BI.

Use of Bayesian Inference in Phylogenetic Analysis

In the Bayesian approach to tree building, the tree topology and the associated model parameters correspond to the hypothesis and the sequence alignments are the data. In a four-taxa scenario, we have three possible unrooted trees (hypotheses), and we need to pick the tree with the highest posterior probability. Thus, the posterior probability of a tree p(T1|D), say tree 1, is

$$p(T1|D)=\frac{p(T1)\times p(D|T1)}{p(T1)\times p(D|T1)+p(T2)\times p(D|T2)+p(T3)\times p(D|T3)}$$

where p (T|D) is the posterior probability of a tree, p (T) is the prior probability of a tree and p (D|T) is the likelihood of a tree.

However, this is an over-simplification because, for a given tree and model of sequence evolution, one can think of different combinations of the parameter values. Each combination of the parameter values can give a slightly different likelihood score for a given tree topology. These parameters include branch lengths, transition matrix and alpha (gamma shape parameter). In the Bayesian framework, we need to explore tree space (different tree topologies) as well as the parameter space for a given tree topology. We will revisit these issues later.

Calculating the posterior probability of a tree is possible when we are dealing with a few taxa, as above, because there are few trees to evaluate. However, as the number of taxa increases, the number of all possible unrooted trees goes up astronomically. This makes the denominator in the above equation difficult to calculate analytically. The denominator, which is the normalising constant, represents probability of the data p(D) and is calculated by summation of p(T) × p(D|T) over all possible trees.

Markov Chain Monte Carlo

To get around this problem, programmes use the Markov chain Monte Carlo approach to estimate the posterior probability of a tree. Monte Carlo approximation is a useful method to estimate the value of the parameter whose true value is difficult to calculate. The Markov process yields the Markov chain, which consists of a finite number of states and transition matrix that provides information regarding the probability of moving from one state to another. The state at time $t + 1$ is dependent only on the state at time t, and not on the state at time t–1. For example, in the case of DNA sequence data, there are four possible states (A, T, G, C) at a site. According to the simplest model of sequence evolution, the Jukes-Cantor model (*see* Chap. 4), the probability of moving from state *i* to state *j* at a site is given by Pij, and the probability of no change is Pii (Fig. 5.25).

Additionally, Jukes-Cantor assumes that all substitutions are equally likely and that Pij = Pji. Based on this information, we can come up with the transition matrix shown in Fig. 5.26.

The transition matrix provides us with information regarding the probability of change of a base to another base, as well as probability of no change, as we move from time *t* to *t + 1*. The values for Pii and Pij can be estimated from the data or may be based on prior information. This matrix can then be used in a Markov chain Monte Carlo (MCMC) simulation to understand sequence evolution, given the Jukes-Cantor model. For example, if we have a sequence with ten bases at time t, we can use MCMC to understand how this sequence will look in, say, *t + 50* generations.

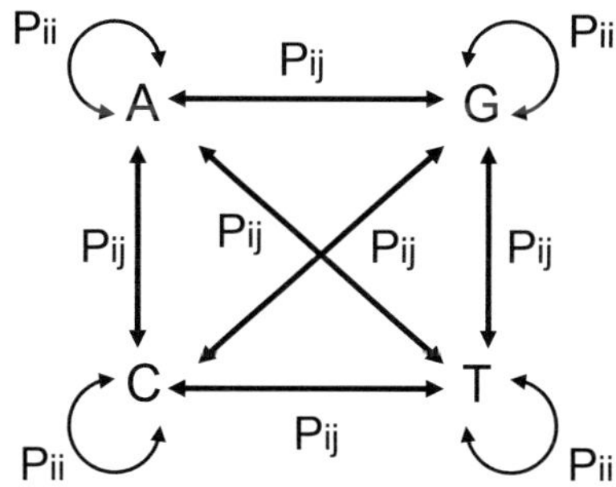

Fig. 5.25 Probability of change Pij and no change Pii

Fig. 5.26 Transition matrix

t \ *t+1*	A	G	C	T
A	Pii	Pij	Pij	Pij
G	Pij	Pii	Pij	Pij
C	Pij	Pij	Pii	Pij
T	Pij	Pij	Pij	Pii

Use of MCMC in Bayesian Inference

In Bayesian inference to tree building, the goal is to determine the posterior probabilities of the trees and pick the tree with the highest posterior probability. However, as mentioned earlier, it is difficult to calculate the posterior probabilities when we have many taxa. Programmes such as *MrBayes* use MCMC to approximate the posterior probabilities of phylogenies by sampling trees from the posterior probability distribution (Huelsenbeck et al. 2002). To do this, the programme starts the Markov chain with a random tree *T* (step 1). In the next generation (link of the chain), a new tree or model parameter value *T′* is proposed (step 2). In step 3, the new tree is accepted or rejected based on the acceptance ratio *R*, where

$$R = \frac{p(T'|D)}{p(T|D)} \times \frac{p(T|T')}{p(T'|T)}$$

p (T′|D) and p (T|D) are the posterior probabilities of trees *T′* and *T*, respectively, p (T′|T) is the probability of proposing *T′* given *T* and p (T|T’) is the probability of proposing *T* given *T′*.

The new tree *T′* is accepted as the current tree in the Markov chain if the acceptance ratio $R \geq 1$ (step 3a).

If $R < 1$, a random number is picked between 0 and 1, and if this number is less than *R*, then the new tree *T′* is accepted as the current tree; otherwise the tree from step 1 (*T*) is retained as the current tree (step 3b). The current tree is then put through steps 2 and 3 for millions of generations to derive the posterior probability distribution.

But how do we calculate posterior probabilities of trees T' and T?

It turns out that this is not a problem, given we are interested in the posterior probability ratio.

$$\frac{p(T'|D)}{p(T|D)} = \frac{p(T') \times p(D|T')/p(D)}{p(T) \times p(D|T)/p(D)}$$

The normalising constant p (D) cancels out, giving us p(D)/p(D)

$$\frac{p(T'|D)}{p(T|D)} = \frac{p(T') \times p(D|T')}{p(T) \times p(D|T)}$$

Thus, we get:

$$R = \frac{p(T')}{p(T)} = \frac{p(D|T')}{p(D|T)} \times \frac{p(T|T')}{p(T'|T)}$$

prior ratio likelihood ratio proposal ratio

To better understand how this works, let us assume that the probability of proposing T' when we are at T equals the probability of proposing T when we are at T' (i.e. Pij = Pji), such that the proposal ratio = 1. Similarly, if we assume that the prior probability of every tree is the same (uniform priors) or that every tree is equally likely to be the correct tree based on prior information, then the prior ratio = 1. This reduces R to just a likelihood ratio of new tree by the old tree, which can be calculated (*see* section on likelihood).

$$R = p(D|T')/p(D|T)$$

This simplification has the following consequences:

When $R > 1$, in every generation, a tree with higher likelihood is picked to represent the current tree in the next generation (step 3a). However, MCMC is set up in such a manner that not only are trees with higher likelihoods picked but also those with the same likelihood score as the earlier tree ($R = 1$, step 3a) or ones with a slightly lower likelihood are picked (step 3b). This is done to ensure that a larger tree and parameter space is explored. If we now plot the number of MCMC generations on the x-axis and the likelihood of the tree on the y-axis, we get the relationship shown in Fig. 5.1. With an increasing number of generations, the likelihood score of the current tree begins to climb and eventually plateaus, a region of the likelihood distribution that is called the stationary distribution. The stationary distribution is the desired posterior probability distribution of interest and the region from which trees are sampled for estimating posterior probability of a tree. To do this, the first part of the distribution where the likelihood scores show an increasing trend are discarded. This part of the distribution is termed as 'burn-in'. The remaining trees, i.e. from the stationary distribution, are summarised to give the Bayesian tree. The posterior probability of a tree is the number of times that the tree topology

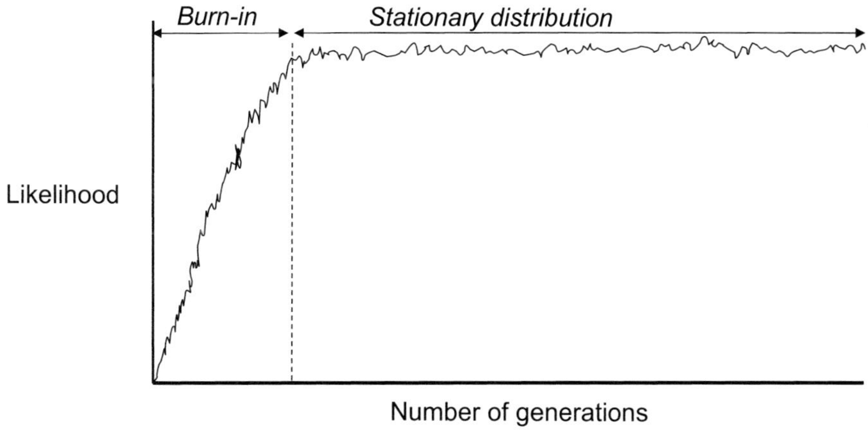

Fig. 5.27 A typical MCMC run, showing an increase in the likelihood score of the sampled tree till it reaches saturation (stationary distribution)

occurs among the sampled trees divided by the total number of trees sampled from the stationary distribution. Similarly, the posterior probability of a node is the number of times that node appears in the sampled trees (Fig. 5.27).

However, just like the other search algorithms, MCMC also can get stuck in a local optimum (see Chap. 6 for discussion on local optima). To get around this problem, programs like MrBayes use Metropolis coupled Markov chain Monte Carlo, (MC)[3]. Here, multiple chains are run, such that a large tree and parameter space is explored. For a detailed treatment of this topic, see Huelsenbeck et al. (2002).

In the introduction, we discussed how the Bayesian approach differs from likelihood in the use of prior information in determining the probability of the hypothesis. With respect to tree search, the Bayesian approach also differs from likelihood (and other optimality-based tree-building methods). Likelihood attempts to find the optimal tree, given a model and known parameter values. Thus, the goal is to reach the highest peak in the tree space. Whereas in the Bayesian approach, the idea is to sample in the vicinity of the peak such that minor variations in topology and parameter values are also taken into consideration. In Bayesian analysis, the final result does not depend on one parameter value but considers all possible parameter values (Holder and Lewis 2003); thus, unlike likelihood, it incorporates the variability of the estimated parameters (Huelsenbeck et al. 2002).

References

Felsenstein J (2003) Inferring Phylogenies. By Sinauer Associates, Sunderland

Holder M, Lewis PO (2003) Phylogeny estimation: traditional and Bayesian approaches. Nat Rev Genet 4:275–284

Huelsenbeck JP, Larget B, Miller RE, Ronquist F (2002) Potential applications and pitfalls of Bayesian inference of phylogeny. Syst Biol 51(5):673–688

Lewis PO (2001) Phylogenetic systematics turns over a new leaf. Trends Ecol Evol 16(1):30–37
Stewart C-B (1993) The powers and pitfalls of parsimony. Nature 361(6413):603–607
Swofford DL (2003) PAUP. Phylogenetic analysis using parsimony (and other methods). Version 4. Sinauer Associates, Sunderland

6 Comparing Different Methods of Tree Building

Abstract

This chapter compares distance (UPGMA, NJ, LS and ME) vs. character-state approaches (ML, MP and BI) to tree building, highlighting their conceptual differences, strengths and weaknesses. Further, the tree-building methods are also compared and contrasted by classifying them into methods that use an algorithmic (UPGMA and NJ) vs. those that employ an optimality criterion (LS, ME, MP, ML and BI) with discussion on computational demands and accuracy. The limitations of clustering methods such as UPGMA that assume clocklike evolution are emphasised. Search algorithms and the problem of local vs. global optima in tree space are explained using a hill climbing analogy. Finally, concatenated vs. coalescent-based methods are introduced, showing how multi-gene data may require species-tree approaches to handle gene-tree discordance. The chapter equips readers with critical insights to select appropriate tree-building methods for their datasets.

Keywords

Optimality criteria · Genetic distance · Molecular clock · Character state · Species tree

The seven different tree-building methods discussed in preceding chapters cover all the major methods of phylogenetic reconstructions (but see coalescent-based tree-building methods). In this chapter, we will discuss how these methods differ from each other, as well as their strengths and weaknesses.

K. P. Karanth, *Molecular Phylogenetics*, https://doi.org/10.1007/978-981-95-9147-3_6

Fig. 6.1 Comparing different tree-building methods. The asterisk indicates that parsimony does not use a model of sequence evolution

Method	Data type	Approach
UPGMA (*Phenetic*)	*Distance*	*Algorithmic*
Neighbor Joining	*Distance*	*Algorithmic*
Least square	*Distance*	*Optimality criterion*
Minimum Evolution	*Distance*	*Optimality criterion*
Parsimony* (*Cladistic*)	*Character state*	*Optimality criterion*
Likelihood	*Character state*	*Optimality criterion*
Bayesian approach	*Character state*	*Optimality criterion*

Clustering (Algorithmic) Versus Search (Optimality Criteria) Methods

Among these methods, unweighted pair group method with arithmetic mean (UPGMA) and parsimony broadly fall under the Phenetic and Cladistic school of systematic, respectively. Overall, these seven methods can be classified into two categories—distance-based tree-building methods and those using character-state approach (Fig. 6.1). As discussed earlier, in distance-based methods, we first calculate pairwise genetic distance between species, and these distances are then used to build the tree. On the other hand, in character-state approaches, a tree is inferred directly from sequence data, i.e. there is no intermediate step of calculating genetic distance. Thus, it is often argued that there is no loss of information as there is no intermediate step where sequence data are converted into genetic distance. More importantly, in the character-state approach, each site in the sequence alignment is considered an independent character and its evolution inferred on the tree separately (*see* introduction to Chap. 5). For these reasons, methods that use character-state approaches are considered better than distance-based methods.

Among the distance-based tree-building methods, UPGMA is problematic, as it assumes a molecular clock. A strict clock assumption is violated in most datasets, and therefore this method can often generate the wrong topology. Thus, it is best to avoid using this method to estimate phylogenetic trees. Neighbour joining (NJ) is a vast improvement over UPGMA as it corrects for rate variation among lineages and usually generates an additive tree. However, NJ does not evaluate all tree topologies (*see* below). In this regard, least square (LS) and minimum evolution (ME) are the preferred distance-based tree-building methods, as they use an optimality criterion. Among the methods using character-state approach, maximum likelihood (ML) and Bayesian inference (BI) are preferred over maximum parsimony (MP) as they use a model of sequence evolution to correct for multiple hits and among site rate variation. However, one can use weighted parsimony to correct for variation in transitions and transversions.

Optimality Criteria-Based Methods: Advantages and Disadvantages

For a given number of taxa, there are multiple possible trees. For example, for a four taxa situation, there are three possible evolutionary trees. Both UPGMA and NJ methods use algorithms to generate a tree, but do not evaluate all possible trees,

Table 6.1 Optimality criteria-based tree-building methods

Method	Optimality criterion
Least square (LS)	LS score
Minimum evolution (ME)	Sum of branch lengths
Maximum parsimony (MP)	Tree length
Maximum likelihood (ML)	Log likelihood
Bayesian inference (BI)	Posterior probability

Table 6.2 Number of all possible unrooted trees for a given number of taxa

Number of taxa	Number of unrooted trees
2	1
3	1
4	3
5	15
6	105
7	954
8	10,395
9	135,135
10	2,027,025
22	3×10^{23}
50	3×10^{74}

whereas the rest of the tree-building methods attempt to evaluate all possible trees and pick the optimal tree based on an optimality criterion. Table 6.1 lists the tree-building methods that use an optimality criterion. These methods optimise the value of the criteria used to pick the optimal tree. For example, LS, ME and MP pick the tree with lowest LS score, sum of branch lengths and tree length, respectively. Whereas ML and BI pick the tree with highest log likelihood and posterior probability, respectively.

Optimality criteria-based methods are far superior to methods that use algorithmic approaches (UPGMA and NJ), because they evaluate all possible evolutionary relationships and therefore are more likely to generate the true tree. These methods work well as long as there are a limited number of taxa in the alignment. However, as the number of taxa goes beyond 12, the number of all possible unrooted trees to be evaluated becomes far too many (Table 6.2).

In the case of optimality criteria-based methods, the programme has to first build all possible trees, and for each of those trees, the value of the chosen optimality criterion has to be calculated. These values have to be compared to pick the optimal tree. Clearly, beyond a certain number of taxa, finding the optimal tree is not possible, as there are far too many trees to evaluate (Table 6.2). This is one of the major issues of optimality criteria-based methods. To get around this problem, tree building programmes use 'heuristic' search approaches to find the best tree, which might not necessarily be the optimal tree. The idea behind the 'heuristic' search is to evaluate as many trees as possible given the limitations of time and computing power. This is achieved by evaluating a subset of trees, from the collection of all possible trees, and the best tree among the subset of trees is picked. The first step in a heuristic search is to generate a random tree or a NJ tree. If ML is the optimality criterion being used, then the likelihood of this starting tree is calculated. This tree is then perturbed, i.e. branches are moved around to generate a new tree. If the new tree has a higher likelihood, then it is accepted as the current tree and the starting tree is discarded (remember in ML we are looking for the tree with highest likelihood score). However, if the new tree has a lower score, then it is discarded and the starting tree becomes the current tree. The current tree is now perturbed again to generate a new tree and the above steps are repeated. The programme goes through millions of iterations of these steps till it reaches a point where there is no further improvement in the likelihood score (best possible score among the sampled trees). Through this process, the programme evaluates millions of trees, but not all the trees. It then picks the best tree it can find among the evaluated trees. However, given that all trees have not been evaluated, there is always a chance that the optimal tree is not present in the subset of trees sampled. The best tree might not be the optimal tree, and there is no way of knowing either way.

Thus, the tree search algorithms can often get stuck in local optima while we are interested in global optima. The concept of local and global optima can be better understood with the hill climbing analogy. In the hill climbing analogy, a person is blindfolded and let loose in a landscape that they are not familiar with. The landscape consists of numerous hills and valleys. The person is then told to climb the highest peak in the landscape. Given the person is not familiar with the landscape and cannot *see* the peaks, how will they find the highest peak?

One approach is to start walking in a random direction from the point of release. If one is walking upwards, then continue walking in that direction. If not, randomly pick another direction and determine if one is walking upwards or downwards. The idea is to constantly correct the direction to ensure one is walking upwards. This will eventually lead us to the highest point in that local area. However, this peak might not be the highest point in the landscape (global optimum). This issue is illustrated in Fig. 6.2, where we have started walking from the base of peak *b* and landed up on a local peak (local optimum). Once on a local optimum, it is not possible to go further upwards as in every direction we move downwards. Thus, we get stuck in a local optimum and have no way of knowing that we have missed the global optimum because we cannot *see* the highest peak. Tree searching algorithms also face a similar problem. Here, the landscape represents the collection of all possible trees (tree space). Each tree is a point in the landscape with a particular likelihood score (altitude). We need to find the tree with the highest likelihood score (global optimum). To avoid getting stuck in a local optimum, tree building programmes run multiple heuristic searches each starting with a different random tree (different points in the tree space). If one runs a large number of independent searches, then it is more likely that one of those searches might reach the global optimum.

With the size of datasets increasing rapidly with every passing year, both with respect to number of taxa and sequence length, faster and more efficient algorithms are needed to generate a reliable tree. With larger datasets, it is important to use a

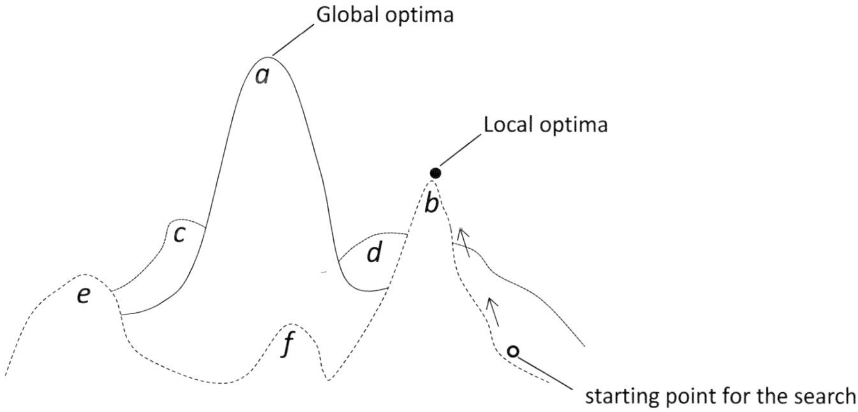

Fig. 6.2 A landscape consisting of peaks (*a–f*) of varying altitudes. The highest peak (*a*) is the global optimum and the remaining peaks (*b–f*) represent the local optima. An open circle is a random point in the landscape that served as our starting point for the search for the highest peak. This search results in the person reaching peak (*b*) (one of the local optima) and not the highest peak (global optimum)

search algorithm that covers a wide tree space and helps us pick the best tree. Thus, while using optimality criteria-based tree-building methods, it is important to pay attention to search settings as these settings are crucial in generating a robust tree.

Concatenated Versus Coalescent-Based Tree-Building Methods

Many of the phylogenetic analysis methods discussed thus far have been based on concatenation of multiple markers and building a tree based on combined data. However, individual markers often have very different evolutionary histories (see Chap. 8). To account for different evolutionary histories of various unlinked genes, which can potentially support different tree topologies, a new class of tree-building methods has gained ascendency. These methods use coalescent-based tree methods that build individual gene genealogies and then generate a tree with the least conflict among markers. It is important to account for the discordance among markers. If each marker is reconstructing a different evolutionary history of the taxonomic group being studied, then it is advisable to use coalescent-based tree methods, also referred to as the species-tree approach (see Chou et al. 2015; Douglas et al. 2022 for details).

References

Chou J, Gupta A, Yaduvanshi S, Davidson R, Nute M, Mirarab S, Warnow T (2015) A comparative study of SVDquartets and other coalescent-based species tree estimation methods. BMC Genomics 16(Suppl 10):S2

Douglas J, Jimenez-Silva CL, Bouckaert R (2022) StarBeast3: adaptive parallelized Bayesian inference under the multispecies coalescent. Syst Biol 71(4):901–916

7 Revisiting Homology: Orthologous Versus Paralogous Genes

Abstract

The chapter revisits the concept of homology with respect to the whole genes as well as a site in a gene. Gene homology due to speciation (orthology) vs. gene duplication (paralogy) are contrasted, with emphasis on why orthologs are crucial for reconstructing species relationships. The challenges posed by nuclear mitochondrial pseudogenes (numts) are explained, highlighting their misleading phylogenetic signals. Broader issues of gene duplication followed by gene deletion resulting in hidden paralogy are discussed. Homology vs. homoplasy at the level of a site in a gene is explored with examples. Convergence, parallel evolution and reversal are introduced as the potential sources of homoplasy. Through case examples, the chapter underscores the importance of critically evaluating sequence datasets for true homology before analysis.

Keywords

Homology · Homoplasy · Orthologs · Paralogs · Convergence

Typically, to generate a species phylogeny, one sequences a particular region of the genome (marker) from all the species of a genus or species group of interest. These sequences, each from a different species, are then aligned, and the alignment is used to generate the phylogeny, using methods described in the preceding chapters. Here it is important that the same region of the genome is being compared across species; in other words, the sequence of the same locus (e.g. a gene) must be generated from different species. These are regions that are similar across species due to their shared ancestry/evolutionary history. Such regions that are similar across genomes of various species are called homologous regions. This is an important aspect of phylogenetic analysis because very often one finds multiple copies of a gene in individuals of a species. In such cases, the same copy of the gene must be used. We explore this issue further, but before that, let us revisit homology.

K. P. Karanth, *Molecular Phylogenetics*, https://doi.org/10.1007/978-981-95-9147-3_7

Homology

Homology is similarity due to common or shared ancestry. Another way to look at homology is that closely related species look similar because they share the same common ancestor or are derived from the same ancestor. Homology can be studied at multiple levels (organism, gene, nucleotide). In the case of homology at the organismal (taxon) level, a good example is that of tetrapods (animals with four appendages or legs). Tetrapods constitute a natural group of vertebrates consisting of amphibians, reptiles, mammals and birds. All members of this group share one important character, i.e. the presence of four appendages. This overall similarity exhibited by these four groups is due to the fact that they share a common ancestor, which was also a tetrapod, and this character was passed on to the daughter lineages (Fig. 7.1). Thus, this similarity is due to homology.

On the other hand, similarity due to convergence, parallel evolution and reversals is called homoplasy (*see* below). For example, bats and birds are also similar in that they both possess wings and can fly (Fig. 7.2). However, this similarity is not due to shared ancestry but due to convergence, i.e. independent acquisition of the same character by distantly related forms. Frogs and ducks share the webbed hind limbs character, but this character evolved independently in these two groups. Thus, the similarity between these two taxa is also a case of homoplasy.

Homologous Genes (Orthologous Versus Paralogous)

At the level of the genes too similarity between two genes can be attributed to homology, i.e. derived from the same ancestor (similarity in gene sequence due to common ancestry). In the example given below, an ancestral species x undergoes diversification, giving rise to four daughter species A–D (Fig. 7.3a). The daughter species acquire the same gene (rectangular box) from their ancestor. When this gene is sequenced from the daughter species and used to build a phylogeny, the species tree depicted in Fig. 7.3b is retrieved.

In the above case, the gene of interest from species A and B is derived from the same gene in the common ancestor y. Similarly, genes from species A to D, all share

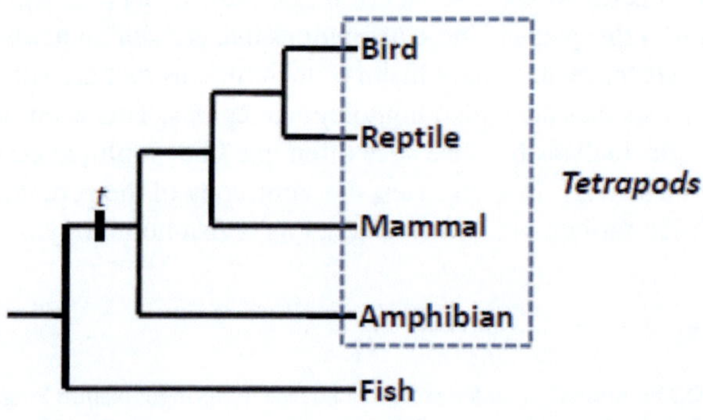

Fig. 7.1 Homology: similarity due to shared ancestry. Amphibians, reptiles, mammals and birds all have four appendages (tetrapods). This character first evolved in the common ancestor of these groups (*t*)

Fig. 7.2 Homoplasy: similarity due to convergence. Birds and bats are similar in that both possess wings (*w*); however, this character (*w*) was independently acquired in unrelated groups

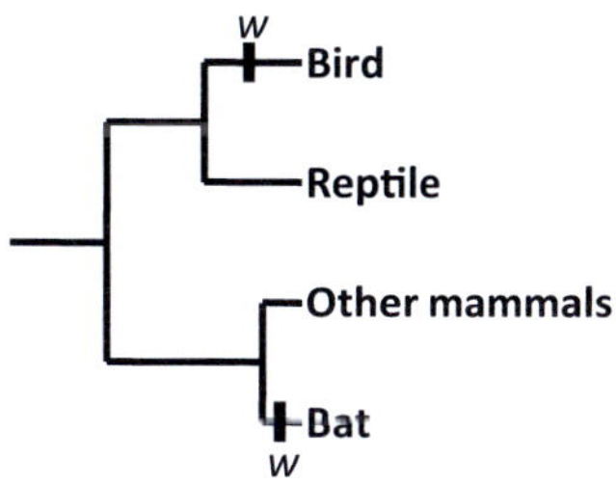

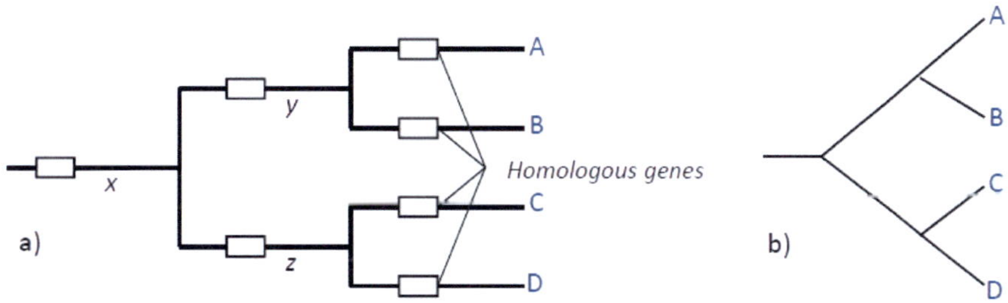

Fig. 7.3 Homologous genes are derived from the same common ancestor. (**a**) x, y and z represent the ancestral species; rectangular boxes represent the same gene in different lineages. (**b**) Phylogeny of the species A–D derived from the gene illustrated in Fig. 7.3a

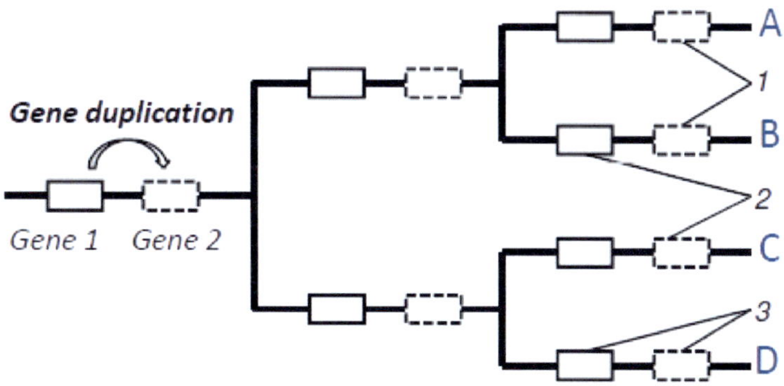

Fig. 7.4 Orthologs (1) and paralogs (2) of homologous genes

the same ancestral gene from species x. Thus, these are homologous genes/regions and are typically present at the same locus in the daughter species A–D. This is a simple and straightforward case of a single-copy gene, but what happens in the case of genes with multiple copies? There are numerous cases of gene duplication events resulting in multiple copies or sometimes families of genes. An example is illustrated in Fig. 7.4.

In this example, the gene has undergone a duplication event in the common ancestor (x) of species A–D. Both copies of this gene (genes 1 and 2) are now present in the daughter species. Thus, in the case of multicopy genes, homologous genes are divided into orthologs and paralogs. Homologous genes separated by speciation are called orthologs, whereas those separated by gene duplication events are called paralogs. We can now compare genes 1 and 2 within and across species (comparisons 1–3). All these comparisons are homologous comparisons; however, in some cases, the two gene copies are being compared within (3) and across species (2), while in the other, the same gene is being compared across species (1). When the same gene (either gene 1 or 2) is compared across species (1), we call it orthologous comparison, and when gene copies (genes 1 and 2) are compared across species (2), we call it paralogous comparison.

In a phylogenetics analysis, care must be taken to ensure that the alignment has the orthologs of a particular gene. If we analyse the two gene copies, discussed above, separately, we retrieve the right phylogeny as both alignments have orthologs (Fig. 7.5).

However, when both the genes are included in the same alignment, the tree shown in Fig. 7.6 is retrieved. In this tree, the two gene copies form separate clusters, each consisting of the four species. The root node of this tree represents the gene duplication event that resulted in the evolution of the two gene copies. Basically, we have taken Fig. 7.4 and stretched apart the two gene phylogenies.

A complication arises if one of the gene copies is deleted in a species. For example, gene 1 is deleted in species C (Fig. 7.7). Thus, species C has just one copy of this gene, while the remaining species have both the copies. The tree generated from an alignment consisting of both the copies is shown in Fig. 7.8a. In this tree, species C is not represented in the gene 1 cluster because this gene was lost on the lineage leading to this species.

When we analyse the two gene copies separately, we retrieve the right phylogeny as shown in Fig. 7.5 given both alignments have orthologs. However, in the phylogeny based on gene 1, species C will be missing. Now let us consider a hypothetical situation wherein we are unaware of the evolutionary history of this gene, and we want to use it to construct the phylogeny of these species. We pick gene 1 based on sequences available from other species groups. We try to PCR amplify and sequence this gene from species A to D. We are able to amplify this gene for species A, B and D, but for species C, the PCR primers pick up the copy, i.e. gene 2.

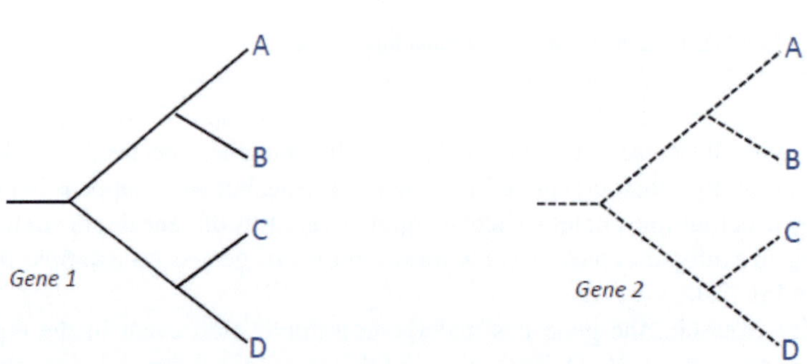

Fig. 7.5 Trees retrieved when the two orthologs are analysed separately

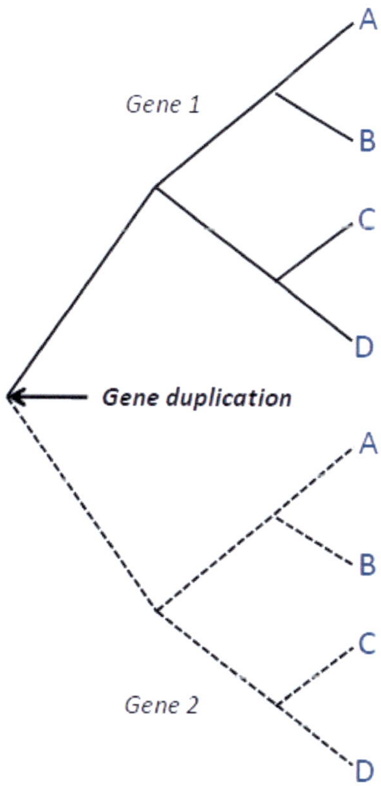

Fig. 7.6 Tree retrieved when the two orthologs are included in the same alignment

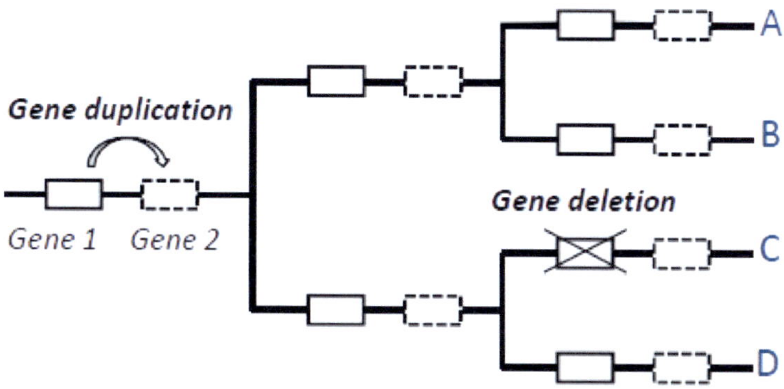

Fig. 7.7 Deletion (cross mark) of gene 1 on the lineage leading to species C

The alignment now has orthologs of gene 1 from species A, B and D, and the paralog of this gene (gene 2) from species C. The resulting phylogeny does not correspond to the species tree (Fig. 7.8b). Thus, it is very important that we use orthologs in our analysis since including a few paralogs can generate erroneous

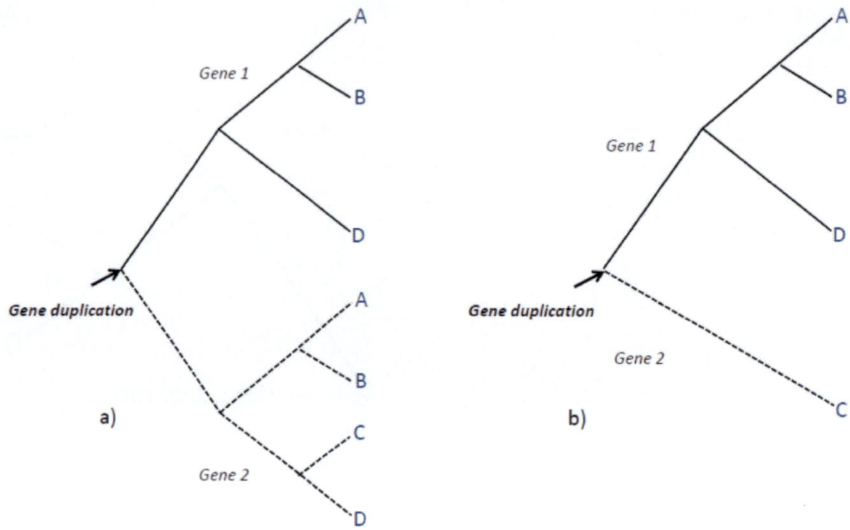

Fig. 7.8 (**a**) Tree retrieved when the two orthologs are included in the same alignment. (**b**) Tree retrieved when the orthologs of gene 1 from species A, B and D are combined with their paralog from species C

phylogenies. It is like using two different genes in the same alignment. This is not an issue in the case of single-copy genes, but when using genes that are part of large gene families, it is very important to establish orthology before accepting the phylogeny and proceeding with interpreting it.

Nuclear Copies of mtDNA Genes

In numerous studies, researchers have reported the presence of nuclear copies of mitochondrial genes in a wide range of plant and animal systems. These nuclear-mitochondrial sequences (numts) arise due to random transfer and incorporation of mtDNA fragments into the nuclear genome. It is widely believed that, after the transfer event, numts lose their function and become nuclear pseudogenes. Thus, numts can be considered 'nuclear paralogs' of mitochondrial sequences and should not be included in an alignment consisting of mitochondrial sequences. Such an alignment will generate an erroneous phylogeny of the taxonomic group of interest. However, this is easier said than done. Typically, studies employing mitochondrial genes use universal primers to amplify and sequence the marker. These primers might sometimes pick the numts instead of the targeted mitochondrial gene. If these sequences go undetected, then they end up in the mitochondrial phylogeny. Thus, numts can be a major issue in phylogenetic studies employing mitochondrial sequences.

How do we detect numts? One of the first places to look for numts is the sequence alignment itself. If the alignment consists of a protein-coding mitochondrial gene, then it is advisable to translate it to generate the protein sequence. In the protein sequence, numts can be identified by the presence of frame shifts due to insertions or deletions (indels) of nucleotides and stop codons that are caused due to nonsense mutations in the DNA sequence. Furthermore, numts often have unique amino acid replacements at sites that are highly conserved in the mtDNA-encoded proteins. This should be done before proceeding to the phylogenetic analysis. Additionally, the tree topology might also hint at the possibility that the alignment consists of numts. Often, in a phylogeny, numts have unusual or unexpected phylogenetic positions. For example, a species might fall in a particular clade in the phylogeny based on nuclear markers, whereas in the mitochondrial tree, the same species might be in a different clade. This incongruence between nuclear and mitochondrial trees should alert the researcher to the possibility of numts in the mitochondrial dataset. However, there are other possible reasons for such incongruence, which are discussed later. In studies employing mitochondrial markers, it is very important that one is aware of the numts issue and that appropriate measures are taken to ensure that true mitochondrial sequences have been used. For a more detailed treatment of this issue, see Karanth (2008) and the references therein.

Homologous Character State

At the level of sequence data, similarity at a site of a gene between sequences can be due to homology or homoplasy. We are familiar with these terms from Chap. 2 where they were discussed with respect to morphological data. Here we will see how these terms can also be used with respect to DNA sequence data. Figure 7.9 shows the phylogeny of six species and an alignment of a gene sequenced from these species. Each site or character in this alignment can attain four possible states: A, T, G and C.

Given the phylogeny, we can now ask the question of whether the similarity in character state between species is due to homology (shared ancestry).

We do this by first assigning the character state to the terminal nodes of the phylogeny for each site. For example, in Fig. 7.10, this exercise has been done for site 1. At this site, species 1 and 2 have the base 'A', and the remaining species have 'C'. Furthermore, using the principle of parsimony, we can deduce that the internal nodes have 'C'. From the phylogeny and the distribution of character states at internal and

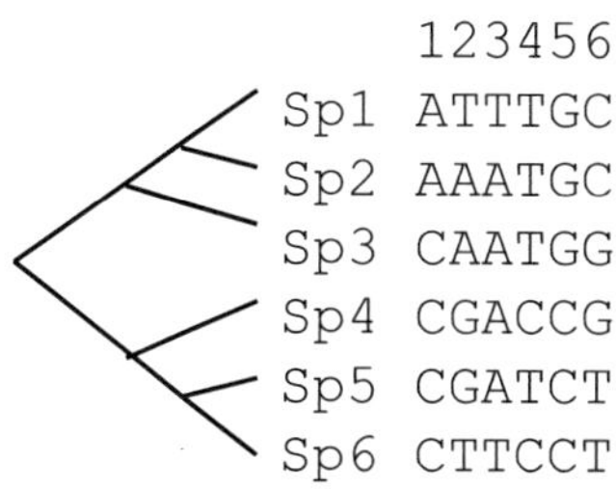

Fig. 7.9 A phylogeny depicting the relationships between six species. Also shown is an alignment of a particular gene sequenced from these species. Numbers 1–6 above the alignment represent a site/character

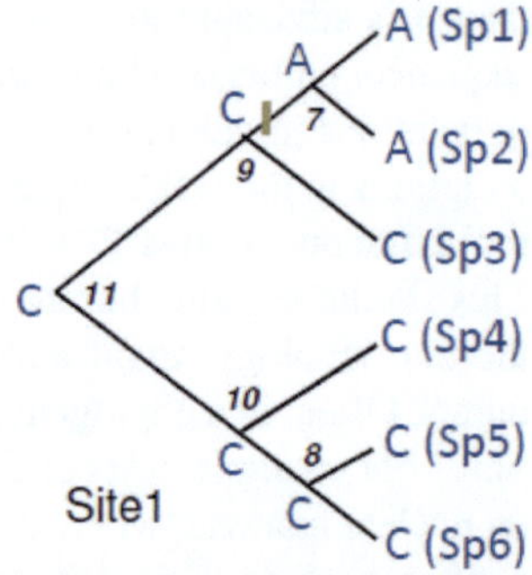

Fig. 7.10 Phylogeny of species 1–6 with character states for site 1 assigned to terminal nodes and presumed character states assigned to internal nodes (7–11)

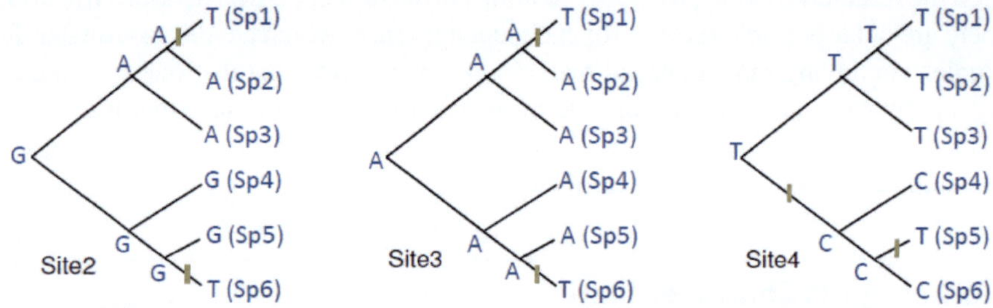

Fig. 7.11 Phylogeny of species 1–6 with character states for sites 2, 3 and 4 assigned to terminal nodes and presumed character states assigned to internal nodes (7–11). Cross marks indicate transition from one state to another along a lineage. Sites 2, 3 and 4 correspond to convergence, parallel evolution and reversal, respectively

terminal nodes, it is apparent that the similarity between species 1 and 2 at this site is due to homology. This is because their immediate common ancestor '7' also had an 'A' at this site. Species 1, 2 and 3 share a common ancestor '9', which had a 'C' at this site. However, there was a transition from C to A along the lineage leading to '7'.

We now contrast homology with homoplasy. Homoplasy is similarity due to convergence, parallel evolution and reversal. For example, in the case of site 2 in Fig. 7.11 species 1 and 6 share the same character state (T). However, these two species are not related in that they do not share an immediate common ancestor. This is convergence as unrelated species have converged to the same character state at this site. Interestingly, the ancestral state at nodes 7 and 8 are different (A and G, respectively). Thus, convergence is independent acquisition of the same character state by unrelated species from different ancestral states. We now move to site 3; here again two unrelated species (1 and 6) have the same character state; however, their ancestral states are similar. This is a case of parallel evolution, which is independent acquisition of the same character state by unrelated species from the same ancestral states. Site 4 is an example of reversal, wherein species 5 has the same character state as species 1–3, but it is more closely related to species 4 and 6. In this case, there has been a reversal of character state from 'C' to a 'T' on the lineage leading to species 5, making it more similar to species 1–3 at this site. Sites that are homoplastic are not consistent with the tree, in that they do not support the tree.

Thus, if the dataset exhibits high homoplasy, then it would support the wrong tree. In our dataset, sites 1, 5 and 6 are consistent with the tree, and in these cases, the similarity that we observe between species is due to homology. See Stewart (1993) for details on homoplasy.

Reference

Stewart C-B (1993) The powers and pitfalls of parsimony. Nature 361(6413):603–607

Gene Tree Versus Species Tree 8

Abstract

This chapter explores the causes of discordance between gene trees and species trees. Processes such as hybridisation, selection, incomplete lineage sorting, sharing of ancestral polymorphism and short internodes are described as sources of conflict. Gene divergence predating species divergence is also explained as one of the reasons that might complicate phylogenetic inference. With respect to incomplete lineage sorting, the link between effective population size of the markers and monophyly is explored. Examples illustrate how single-locus analyses can misrepresent species histories. The chapter emphasises multi-locus and coalescent-based approaches as solutions, urging caution when interpreting individual gene trees. It provides a conceptual framework for understanding why phylogenetic trees can differ across markers and how to reconcile them to recover accurate species histories.

Keywords

Discordance · Deep coalescence · Hybridisation · Selection · Gene genealogy

One of the fundamental questions in phylogenetics is whether gene trees always correspond to species trees. Alternatively one might ask the question 'Are trees generated using two different genes always congruent with each other?' These are important issues because we are estimating the species phylogeny based on gene trees. At the onset, it must be pointed out that genes, given they are embedded in a species, are expected to track the evolutionary trajectory followed by the species. Thus, it is reasonable to assume that gene trees will always be congruent with species trees as long as the alignment has orthologous sequences (Chap. 7). For example, in Fig. 8.1, the gene phylogenies are congruent with species phylogeny in that every time there is a speciation event (bold arrows in the species tree), there is also a corresponding split in the gene trees (arrows in the gene trees).

However, under certain conditions, gene trees might not be concordant with the species tree. In this section, we explore these conditions.

K. P. Karanth, *Molecular Phylogenetics*, https://doi.org/10.1007/978-981-95-9147-3_8

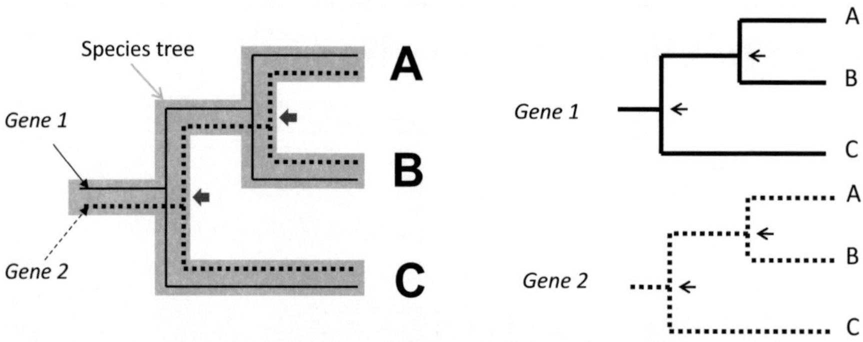

Fig. 8.1 Gene trees typically track species trees. Grey bar indicates the species tree and the dotted and undotted lines indicate gene phylogenies. Arrows indicate speciation events

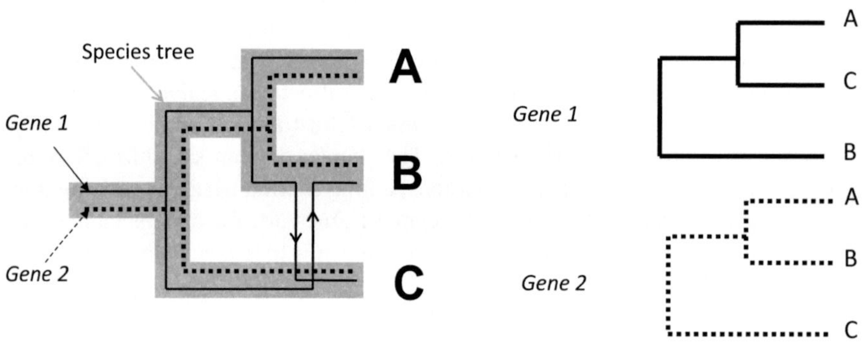

Fig. 8.2 Phylogenies generated using two different genes are not concordant due to hybridisation between species B and C. Grey area represents species tree

Hybridisation

One of the simplest explanations for lack of concordance between two gene trees is hybridisation. Hybridisation might be recent, between closely related species, or might have happened in the distant past (ancient hybridisation). Hybridisation introduces alleles from one species into another species, and this process is called introgression. Not all genes might introgress across a hybrid zone. For example, in Fig. 8.2, gene 1 has introgressed across the species boundary in both directions, whereas gene 2 has not. Thus, the phylogeny based on gene 1 is not concordant with that of gene 2. Researchers are often frustrated when gene phylogenies are not concordant with each other and begin to doubt their data or sequence analysis. Such discordance should not be viewed as problematic, as it is telling us something about the evolutionary history of the species. In this case, it is telling us that there might be ongoing hybridisation between species B and C. For example, Karanth (2008,

2010) noted that disagreement among markers with respect to the phylogenetic position of capped and golden langurs was suggestive of the hybrid origin of this lineage. This scenario should be explored further with additional molecular data as well as ecological and behavioural data from the field.

In our example, we know the species phylogeny, and therefore we can conclude that gene 2 phylogeny is concordant with species phylogeny and that gene 1 has introgressed. A priori, how do we know which gene phylogeny is giving us the right species phylogeny? To address this question, one has to use many unlinked markers and also different kinds of markers, for example, X and Y chromosome markers, autosomal genes, mitochondrial genes, etc. (also *see* FAQs). Among microorganisms, horizontal gene transfer is akin to hybridisation of higher organisms and will result in discordance between gene trees.

Selection

Typically, all protein-coding genes are under selection, but they are usually under purifying or negative selection. Negative selection occurs when a mutation that alters the amino acid sequence is purged from the population. Thus, all non-synonymous changes, i.e. nucleotide substitutions that cause amino acid replacement, are not tolerated, whereas synonymous or silent changes are tolerated. Negative selection maintains the protein sequence and thus its structure and function. It is an important process through which deleterious mutations are removed from the populations (hence also called purifying selection).

Take the example illustrated in Fig. 8.3. Here a particular gene has been sequenced from four different species. The protein sequence coded by this gene is identical across species (conserved) and is given at the bottom of the alignment. However, the DNA sequences of this gene from the four species show considerable variation. This is because all of the substitutions in the DNA alignment are silent/synonymous changes, or changes that do not alter the protein. Synonymous substitutions are largely confined to third position of the codon. Third position silent substitutions are also called neutral substitutions as they escape selection and are free to change. Nevertheless, the gene as a whole is under (negative) selection as the amino acid sequence is conserved across species.

However, in cases where a gene has multiple copies, the duplicated gene copy is not under any selection constraint as the original gene continues to perform the desired function. Such genes can accumulate non-synonymous changes, i.e. nucleotide substitutions that result in a change in an amino acid and therefore code for a different protein. If this new protein is beneficial to the organism, then these non-synonymous changes will be selected. Genes wherein non-synonymous substitutions outnumber synonymous substitutions are said to be under positive selection. Usually, genes under positive selection have been recruited for a new function. These genes evolve rapidly, and their sequences are very different from their orthologs in other species. If a gene chosen for building a phylogeny is under positive selection in one or more species of the ingroup, then such a gene might generate an

```
                *   *       *   * *
Species 1 ATG GGC ACT TAT CCA ATT TTA TAA 24
Species 2 ATG GGT ACT TAT CCC ATC TTA TAA 24
Species 3 ATG GGT ACA TAT CCT ATC CTA TAA 24
Species 4 ATG GGC ACG TAT CCT ATT CTA TAA 24
          Met Gly Thr Tyr Pro Ile Leu Stop 8
```

Fig. 8.3 Nucleotide sequences of a gene from four different species with the corresponding amino acid sequence of the protein coded by this gene

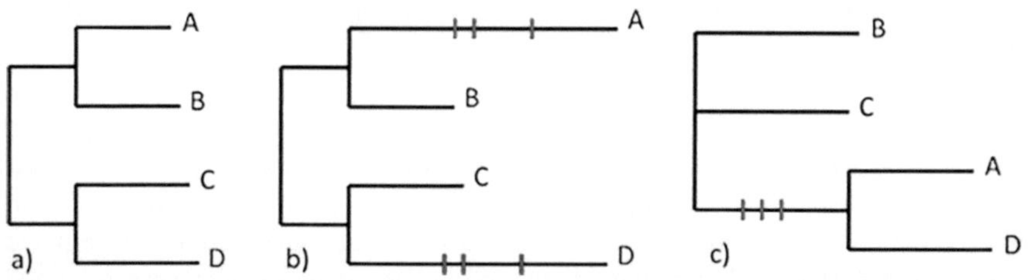

Fig. 8.4 (**a**) Phylogeny of a species group consisting of four species (A, B, C and D). (**b**) Expected phylogeny of this species group for a gene under positive selection in species A and D. Vertical bars represent similar patterns of nucleotide substitutions in species A and D. (**c**) Phylogeny based on the gene under positive selection

erroneous phylogeny and must not be used. Take the example illustrated in Fig. 8.4. Figure 8.4a shows the phylogeny of a species group consisting of four species A, B, C and D. This is the tree that will be generated from most genes sequenced from these four species. In Fig. 8.4b, the expected phylogeny of these species for a gene that is under positive selection in species A and D is illustrated. In this phylogeny, the lineages leading to species A and B are unusually long due to rapid accumulation of substitutions. However, when the gene under positive selection is used to derive the species phylogeny, an erroneous tree illustrated in Fig. 8.4c is generated. This is because this gene is under similar selection pressure in species A and D and has experienced similar patterns of nucleotide substitutions (vertical bars) in these two unrelated species. Thus, the similarity in sequences between species A and D is due to homoplasy (convergence to be precise) and not due to homology.

A priori, how do we know that a gene is under positive selection in one or more species? The first step is to look at the amino acid sequences. If the sequence for a particular species is very different from the orthologs from other species, then we might suspect positive selection. Secondly, one must use multiple genes for generating species phylogenies. Most genes will generate similar tree topology; however, one particular gene might generate a topology very different from the rest. This should alert us to the possibility of this particular gene being under positive selection. One of the classic examples of how positive selection influences phylogeny comes from the lysozyme gene. This gene has undergone convergent evolution in cows and langur monkeys, in that it has been recruited for a similar function in these two distantly related taxa. In the lysozyme gene phylogeny, langur monkeys branch with cows and not the primates—a relationship that is not consistent with the known phylogeny (Stewart et al. 1987). This phenomenon is often referred to as long branch attraction, i.e. species with long branches due to a higher rate of nucleotide substitutions tend to branch together. However, it must be noted that higher rates of nucleotide substitutions can also occur due to processes other than positive selection, such as short generation time.

Shared Ancestral Polymorphisms in Recently Diverged Species

Typically, during speciation, the subpopulations of an ancestral species get isolated from each other due to a barrier, and these subpopulations evolve into separate species. For example, a mountain range might emerge in the middle of a species' range, dividing the population into two subpopulations. These subpopulations eventually diverge from each other due to a combination of factors such as lack of geneflow between them, drift and selection. This process leads to speciation, and the end product is two daughter species. The speciation process, in turn, goes through different stages of divergence. Early in the speciation process, the incipient daughter species share alleles (polymorphisms) at multiple loci, a phenomenon called incomplete lineage sorting (ILS) and therefore show very little divergence between them. With time, species-specific alleles get fixed in the daughter species, and the two entities become genetically distinct from each other.

Take the example illustrated in Fig. 8.5 wherein we have two species pairs at different stages of evolutionary divergence. The daughter species within each of these pairs have been isolated from each other for different time periods. Species A and B have undergone speciation recently, whereas species D and C underwent speciation in the distant past. Given this scenario, if we now sequence a gene from multiple individuals per species (A, B, C and D) and generate a phylogeny, what would the

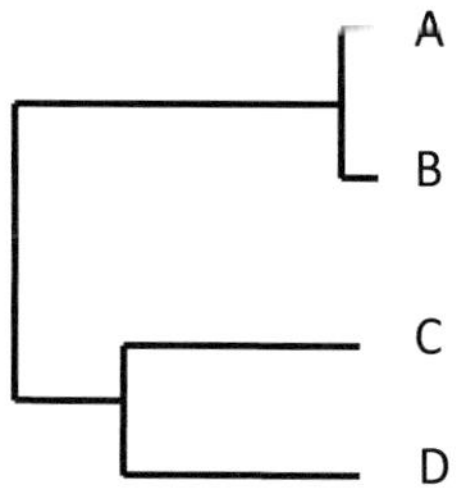

Fig. 8.5 Two species pairs at different stages of evolutionary divergence. Species A and B have separated from each other recently as illustrated by the 'shallow depth' between them, whereas species C and D have separated from each other in the distant past

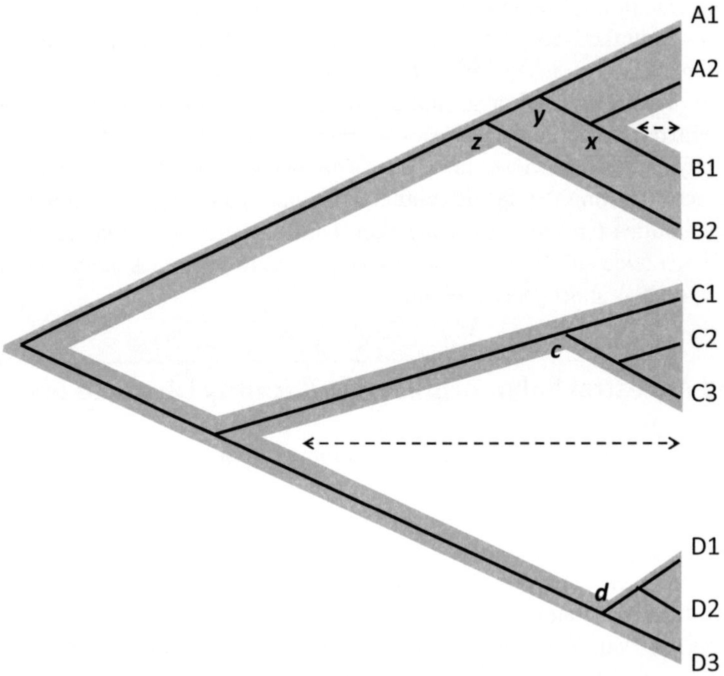

Fig. 8.6 Two species pairs in different stages of evolutionary divergence. Filled-in grey area represents species phylogeny, and black line represents gene phylogeny. Dashed lines depict time since the two species pair separated from each other. Letters at the internal nodes represent common ancestors in the gene phylogeny

topology look like? Such a phylogeny is depicted in Fig. 8.6. In the case of the older species pair (C–D), two distinct clusters are retrieved, each corresponding to the members of the two species. This means that a member of species C (say C1) is more closely related to its conspecific (C2 and C3) than it is to a heterospecific (say D1). The same is true for species D.

Typically, members of any species are expected to be more closely related to each other than to members of a sister species. For example, humans and chimps are sister species; if you pick any two human individuals, they will be genetically more closely related to each other than either of them is to a chimp. The technical term we use to describe this condition is monophyly, i.e. species C is monophyletic and so is species D. Together, species C and D are reciprocally monophyletic. Members of a monophyletic group stem from the same lineage or common ancestor, *c* and *d* for C1–3 and D1–3 respectively.

However, species monophyly is not always a norm, particularly in the case of closely related and recently diverged species due to ILS. For example, in the case of the species pair A–B, members of the two species do not form separate clusters. Instead, B1 is more closely related to A2 than B2. This is because A and B have recently diverged and still share alleles (polymorphisms) at this locus. In our example, A2 and B1 share the same allele at this locus and are therefore sister to each other in the phylogeny. The condition wherein members of sister species do not fall in separate groups/clusters is called polyphyly. In our example, we say species A and B are polyphyletic with respect to each other. This scenario is also referred to as incomplete lineage sorting. With time, the two species will attain reciprocal monophyly due to various evolutionary forces such as lack of geneflow, drift and selection. Simulations have shown that time to reciprocal monophyly takes 4Ne generations, wherein Ne is the effective population size of the species. Thus, in the case of recently diverged species, the gene phylogenies might not retrieve the individual species as monophyletic units. See Avise (1994) for a detailed treatment of this topic.

Lack of species monophyly can confound species relationships when we have three or more closely related and recently diverged species. This is because different genes might retrieve different phylogenies due to the stochastic process of allele sharing/lineage sorting between these species. This scenario is illustrated in Fig. 8.7. In the species phylogeny, A + B are sister to C. However, all their gene genealogies do not support monophyly because they have diverged from each other recently. When we estimate the phylogeny of these species, the topology retrieved will depend on the gene used, or the individual sample used in the analysis. For example, when we sequence gene 1 from individuals A2, B2 and C2, we get the right species phylogeny (Fig. 8.7a). However, when the same gene is sequenced from A1, B1 and C1, we retrieve a different tree. Similarly, when a different gene is used (Fig. 8.7b), once again a new topology is retrieved. Thus, between these two genes, three topologies are retrieved; A + B sister to C, A + C sister to B and B + C sister to A.

Which among these three topologies is the correct species tree? There is no easy answer to this question. One approach is to present an unresolved tree wherein all the three species emanate from the same node (trichotomy). This scenario might be reasonable, given the three species have diverged from each other recently, and therefore sufficient time has not elapsed for clear-cut relationships to be established.

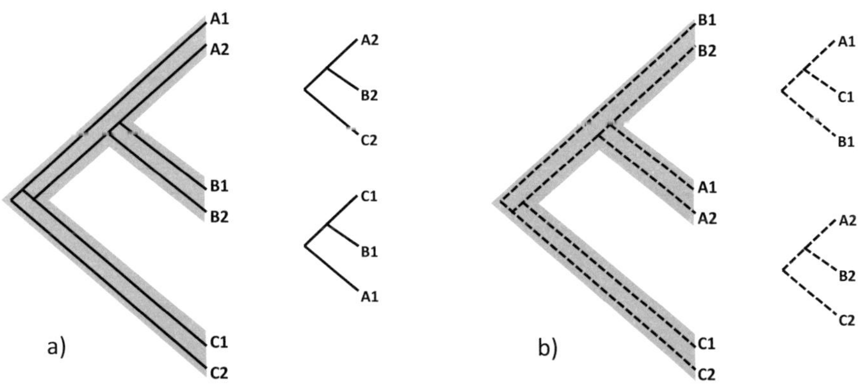

Fig. 8.7 Species phylogenies with gene genealogies for two genes (**a** and **b**) mapped onto them. Also shown are the corresponding gene phylogenies when different individuals are sampled

Perhaps we have to wait for a couple of million years for monophyly. Alternatively, a majority rule can be applied to generate a species phylogeny. For example, for a given gene, sequences are generated from at least two individuals per species. These individuals should be collected from as far apart from each other as possible. For this gene, we then derive phylogenies for all possible combinations of individuals from different species. In our example, we generate trees for these combinations: A1, B1, C1 and A2, B2, C2; however, other combinations are possible such as A1, B1, C2; A1, B2, C2; A1, B2, C1; A2, B1, C2; A2, B2, C1 and so on. We can then pick the most common topology as the species phylogeny. We can implement this approach across multiple genes. Alternatively, species tree/coalescent approach can be used (see page 104).

Short Internodal Time

The time between two speciation events is an important factor that influences the topology of the tree or support for a particular node. Take the example illustrated in Fig. 8.8a. In this phylogeny, two speciation events are depicted resulting in the evolution of three species (A, B and C). The first speciation event was at the node (1) representing the common ancestor of species A, B and C. This speciation event resulted in two species: one leading to species C and the other to the common ancestor of species A and B. The second speciation event occurred at the node (2) representing the common ancestor of species A and B. This speciation event led to the evolution of species A and B. The time (*t*) between these two nodes (1 and 2) represents the time between the two speciation events. Ideally, adequate time should elapse between these two speciation events for sufficient phylogenetically informative characters to accumulate such that a well-resolved tree is generated. In the example discussed, the horizontal bars represent the phylogenetically informative characters (synapomorphies) that support the sister relationship of species A and B

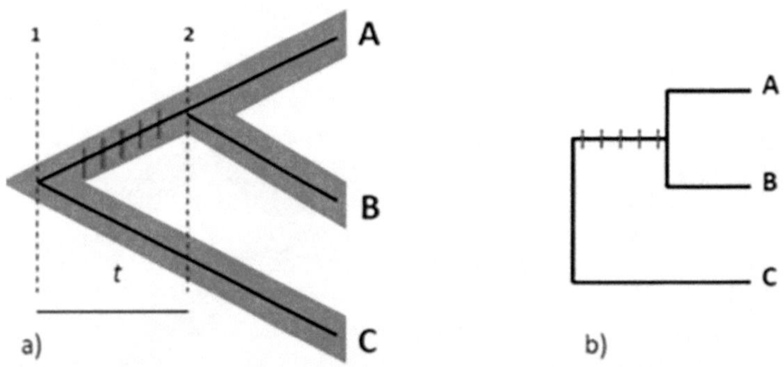

Fig. 8.8 Phylogeny of three species with long internodal time. (**a**) The numbers 1 and 2 represent the two speciation events (nodes) and *t* is the time between speciation events. (**b**) Time *t* can be absolute time between speciation events or the branch length between nodes 1 and 2. Vertical bars represent characters that support sister relationship of species A and B

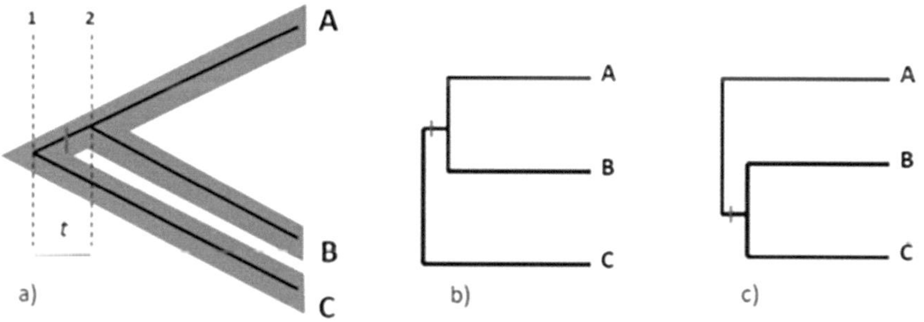

Fig. 8.9 (**a**) Phylogeny of three species with short internodal time. (**b**) The right phylogeny of the three species based on a gene. (**c**) Incorrect phylogeny generated due to back mutation of a phylogenetically informative character

(Fig. 8.8a). In this case, sufficient changes have accumulated along this lineage, and we get a well-supported tree (Fig. 8.8b).

However, if the time between the two speciation events is very short (Fig. 8.9a), then very few characters that support the sister relationship between A and B accumulate between nodes 1 and 2. One is alerted to this scenario by a short branch length between nodes 1 and 2 (Fig. 8.9b) and very low bootstrap support at node 2. Nevertheless, we are still in the safe zone as the right tree has been retrieved. The problem arises when the few characters that support the sister relationship between A and B undergo subsequent change, thereby confounding the actual relationship. For example, in Fig. 8.9c, this character has undergone a substitution in species B, making it more similar to species C. This is never an issue when many characters support a particular relationship (as in Fig. 8.8), as a few back mutations will not affect the topology. Thus, short intermodal time can also result in conflicting topologies across genes. Some genes will generate the right phylogeny, while other more rapidly evolving genes might generate incorrect trees due to reversal of characters or back mutations.

It must be noted that the issue of short internodal time is usually confined to certain parts of a larger phylogeny. For a given phylogeny, it is very unlikely that short internodal time will be an issue for all the internal branches.

Gene Divergence Precedes Species Divergence

One often assumes that gene divergence occurs concomitantly with species divergence. In other words, as populations get isolated and begin to evolve into separate lineages (species), the genes embedded in these populations also begin to diverge due to accumulation of mutations and lack of geneflow (Fig. 8.8). This is a reasonable assumption for most genes. However, for some genes with much ancestral polymorphism, this might not be true, as illustrated in Fig. 8.10. Here, the gene in question has three alleles in the common ancestor of species A, B and C. These alleles have diverged from each other before the three species diverged from each

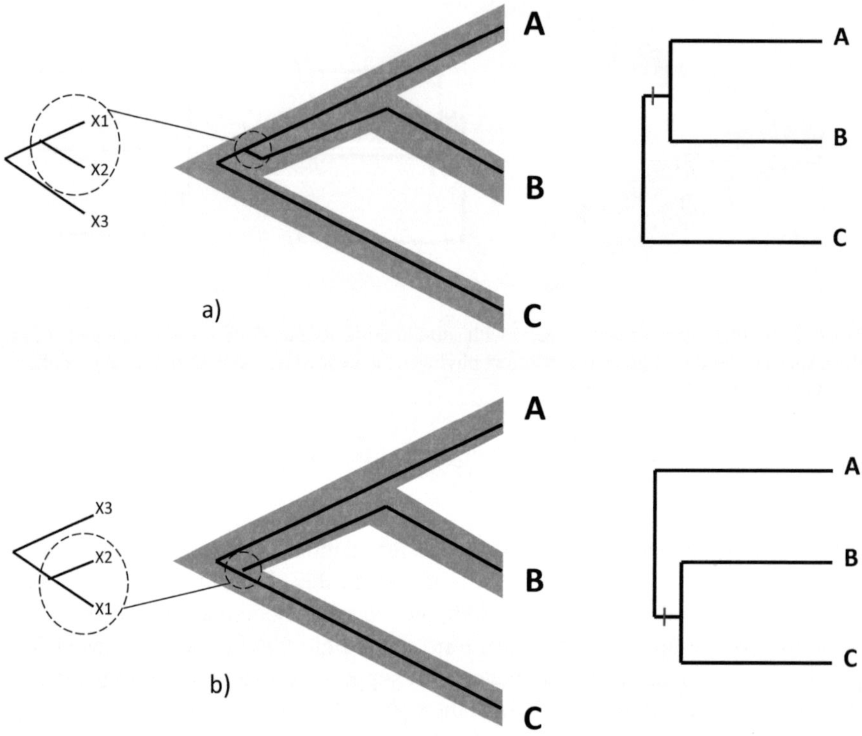

Fig. 8.10 Phylogeny of three species where gene divergence has preceded species divergence. (**a**) The right phylogeny of the three species is retrieved when species A and B have alleles X1 and X2, respectively. (**b**) Incorrect phylogeny is generated when species A and B have alleles X3 and X2, respectively. Relationships between the alleles are shown in the slant cladograms to the left of the species trees

other. Thus, the common ancestor is polymorphic at this locus. The three alleles have a certain relationship, i.e. alleles X1 + X2 are sister to X3. In the Newick format, it would be (X1, X2), X3. Given this relationship between the alleles, the topology of the tree derived from this tree will depend on how these alleles sort in the daughter species.

For example, if alleles X1 and X2, which are sister to each other, land up in species A and B (Fig. 8.10a), then the right phylogeny is retrieved (Fig. 8.10b). However, it must be noted that the genetic distance between species A and B in Fig. 8.10a reflects the divergences between alleles X1 and X2 and not species divergence. On the other hand, these alleles might sort differently, as shown in Fig. 8.10b. The tree topology in this case depends on how these alleles sort themselves in the daughter species. Thus, if gene divergence precedes species divergence, then the ancestral species is polymorphic at that locus, and the random sorting of this polymorphism (alleles) can generate incongruence between species and gene trees (Graur and Li 2000). This phenomenon is also referred to as deep coalescence (Maddison 1997).

Effective Population Size of the Chosen Marker

Thus far, we have discussed why, sometimes, trees generated by two different genes might not be concordant and under what conditions gene trees and the expected species tree might not be congruent. In this regard, another complicating factor is the effective population size of the chosen gene (marker) which, in turn, influences time to reciprocal monophyly (*see* above). The effective population size of a gene depends on whether the gene is present in the nucleus or in the cell organelle and/or the social organisation of the animal/plant system. For example, in a mammalian system, the chosen marker could be mitochondrial or nuclear. Within the nuclear genome, we have autosomal and sex chromosomal markers. Mitochondrial DNA is haploid and maternally inherited (from mother to daughter), and therefore the effective population size of mtDNA marker is one fourth that of nuclear autosomal genes. Thus mitochondrial markers are predicted to attain reciprocal monophyly four times faster than nuclear autosomal markers (Moore 1995). In recently diverged species, mtDNA markers often attain reciprocal monophyly, whereas nuclear markers retrieve a polyphyletic tree. This difference in topology reflects the difference in the nature of the markers used. In plant systems, in addition to mitochondrial and nuclear markers, we have chloroplast markers which also exhibit uniparental inheritance and are haploid.

The social organisation of the animal/plant group under study might also influence time to monophyly. The above contrast between nuclear and mtDNA markers assumes that both males and females of a species disperse and to the same extent. However, we know that some primate species exhibit female natal philopatry (females stay back in the territory they were born in). In this scenario, monophyly in the mtDNA marker will be delayed due to lack of admixture of mitochondrial haplotypes within the species. Y-chromosomal markers are very similar to mtDNA markers with respect to effective population size, because they are also haploid and exhibit uniparental inheritance (from father to son). Mitochondrial markers inform us about female-mediated geneflow, whereas Y-chromosomal markers male-mediated geneflow. Here again, time to monophyly in the case of the Y marker is predicted to be four times faster than nuclear autosomal markers. However, in the case of both mtDNA and Y markers, we assume that the sex ratio is 1:1; more specifically, the effective population size of males and females in the population is the same. This assumption is often violated in species where the variance in reproductive success of the males is very high. In other words, very few males reproduce and most males do not leave behind offspring in the next generation. Under such a situation, the effective population size of the Y marker is further reduced, in turn influencing time to monophyly. See Chap. 12, the section on 'What kind of markers to use?' for more discussion on this topic.

References

Avise JC (1994) Molecular markers, natural history, and evolution. Chapman & Hall, New York

Graur D, Li W-H (2000) Fundamentals of molecular evolution, 2nd edn. Sinauer Associates Inc., Sunderland

Karanth KP (2008) Primate Numts and reticulate evolutions of capped and golden leaf monkeys (primates: Colobinae). J Biosci 33(5):761–770

Karanth KP (2010) Molecular systematics and conservation of the langurs and leaf monkeys of South Asia. J Genet 89(4):393–399

Maddison WP (1997) Gene trees in species trees. Syst Biol 46(3):523–536

Moore WS (1995) Inferring phylogenies from mtDNA variation: mitochondrial-gene trees versus nuclear-gene trees. Evolution 49(4):718–726

Stewart CB, Schilling J, Wilson A (1987) Adaptive evolution in the stomach lysozymes of foregut fermenters. Nature 330:401–404

Statistical Tests

9

Abstract

This chapter covers statistical methods for evaluating tree robustness. Bootstrapping is introduced as a resampling technique to assess confidence in clades. The logic, procedure and interpretation of bootstrap values are explained, with emphasis on their limitations. Hypothesis testing in phylogenetics is discussed through likelihood ratio tests, alternative topological tests and model selection. The chapter highlights how different hypotheses can be tested in a phylogenetic framework. By integrating statistics into phylogenetics, it underscores the importance of measuring uncertainty rather than assuming trees as absolute truths.

Keywords

Bootstrapping · Hypothesis testing · Model selection · Likelihood ratio test · SH test

Bootstrapping

A phylogenetic tree tells us how species are related to each other and also the degree of relatedness between sister species or groups. However, often, we are interested in knowing how well the data support the positions of various internal nodes. This is because some nodes might be better supported by the data than others. To better understand this scenario, take the example illustrated in Fig. 9.1. Here, the genetic distance between species 1 and 2 as well as between 3 and 4 is the same (p-distance = 1/9). Nevertheless, there are four synapomorphies that unite species 1 and 2 (sites 2, 4, 6, 8), whereas in case of species 3 and 4, only one synapomorphy (site 5) unites them. Thus, when we look at parsimony-informative sites, there is more support for sister relationship between species 1 and 2 than for species 3 and 4. This information is not captured by the phylogram generated using this dataset. The only hint of this information is in the long branch leading to species 1 and 2. But how do we quantify it?

K. P. Karanth, *Molecular Phylogenetics*, https://doi.org/10.1007/978-981-95-9147-3_9

Fig. 9.1 Hypothetical dataset and the corresponding parsimony phylogram

```
    123456789
Sp1 AGGTTATTG
Sp2 AGGTGATTG
Sp3 ACGACGTAG
Sp4 ACGACGTGG
Sp5 ACTAGGCCC
```

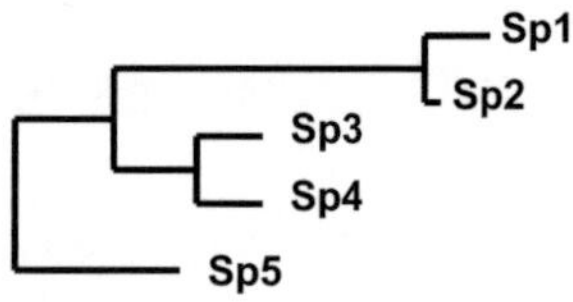

Tree building programmes use bootstrapping to quantify the support for various nodes. Bootstrapping is a simple technique to test the relative stability of groups within a phylogenetic tree. It is a measure of reliability of a particular node or how much support there is for a node in the dataset. To better understand bootstrapping, let us consider the following scenario. If we resampled the five species in our example and reconstructed the phylogeny, would we retrieve the same tree? To do this, the five species are resampled from the field, taking care to sample a different individual of each species. The same marker is then sequenced from these individuals, thereby generating a new dataset. A tree is then built using the new dataset. We can do this, say, 100 times, resulting in 100 datasets and 100 trees. Ideally, the same tree topology should be retrieved from the 100 datasets, which would mean that there is 100% support for the various internal nodes. However, some trees among these 100 trees might have slightly different topologies. This is because of intraspecific variation, i.e. the sequences for a given species from different individuals might not be identical across datasets. Typically, interspecific (between species) variation is far greater than intraspecific (within species) variation. However, in the case of multiple closely related and recently diverged species, inter- and intraspecific variations across species might overlap (see Chap. 8). This scenario, in turn, might give us different topologies depending on the exemplar specimen chosen to represent a species. Bootstrapping is a way to quantify this variation. High bootstrap values can be interpreted as low intraspecific variation when compared to interspecific variation. Similarly, short internodal time between two speciation events (see Chap. 8) can also give us low support for certain nodes due to lack of synapomorphies uniting these lineages.

Deriving bootstrap values by resampling from nature is not practical due to limitations in time and resources. Additionally, in some cases, the species of interest might be endangered, and therefore collection of hundreds of individuals is not advisable. Instead of resampling from nature, one can resample, with replacement, from the original data to derive multiple bootstrap replicates (Fig. 9.2).

To generate a bootstrap dataset, the programme randomly picks one character from the original dataset. A character here represents a column from the original alignment consisting of the corresponding bases for each species at a particular site. For example, in bootstrap replicate (BR) 1, the programme has randomly picked character 9 to start populating the first bootstrap dataset (Fig. 9.2). Character 9 is not removed from the original dataset and can be picked again; hence, we call it sampling with replacement. The programme then randomly picks another character from the original dataset, with replacement, and this process continues till BR1 has nine characters as in the original dataset. Thus, for a given bootstrap dataset, the

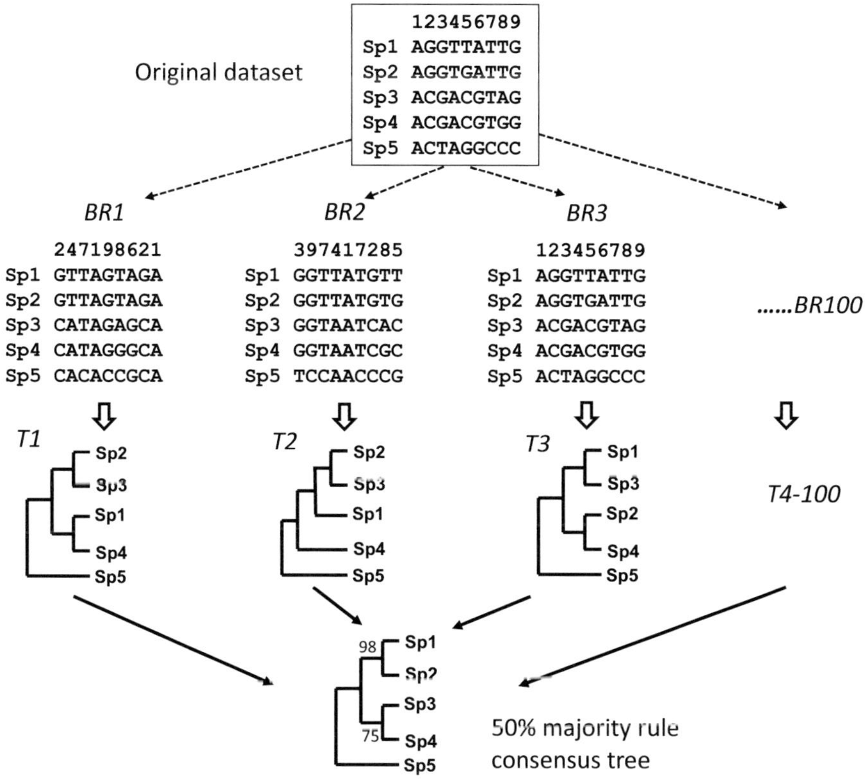

Fig. 9.2 Diagram depicting the various steps in bootstrapping. BR stands for bootstrap replicate, BR1–100 represent the bootstrap datasets generated by resampling from the original dataset. T1–100 are the individual trees derived from these datasets. These trees are used to construct the consensus tree

programme might pick the same character multiple times, which is a consequence of sampling with replacement. Additionally, the order in which these characters appear in the bootstrap dataset is also very different from the original dataset. In this manner, the programme generates many datasets or bootstrap replicates. In our example, we have generated 100 bootstrap replicates. For each bootstrap dataset or replicate, the programme generates a tree based on the chosen tree-building method. These trees are then collated to generate a consensus tree. A consensus tree is the tree that has been retrieved in most of the replicates. The numbers at the nodes of the consensus tree are the bootstrap support for that node (Fig. 9.2). These numbers represent the number of trees in which the two daughter lineages have been retrieved as sister taxa among the 100 trees. For example, species 4 and 5 were sister to each other in 70% of the trees. Often programmes report only 50% bootstrap support, and we call it 50% majority rule consensus tree. The nodes with <50% support are either collapsed or they are retained and the support not reported. Usually, >70% bootstrap values are interpreted as good support, but ideally one should aim for >90% support.

Bootstrapping is an important and useful tool for describing trees and must be reported at all times. An important aspect of bootstrapping is the number of replicates. One should aim for higher numbers of replicates for robust bootstrap values. This is particularly important in datasets where the sequence lengths are large. A greater number of replicates would ensure that every site has been picked at least once in one of the bootstrap datasets. Typically, the lower limit for the number of replicates is 1000, and the upper limit depends on computing power and time. However, in the case of optimality criteria-based tree-building methods, another complication arises. Here, the number of taxa can have serious implications on how many replications one can realistically undertake (see Chap. 6). Bootstrapping can be implemented for all tree-building methods except Bayesian inference. In the case of the Bayesian approach, the posterior probability of a node is calculated by summarising the trees in the stationary distribution.

Hypothesis Testing

Phylogenetic trees built using molecular data often do not agree with morphology-based taxonomy. In such situations, one would like to know if the molecular tree is very different from the morphology-based tree or if the two trees are broadly congruent. To understand this issue, let us consider an example where we are interested in determining the evolutionary relationship between six species currently assigned to two different genera. Genus X has three species X1, X2 and X3; similarly, genus Y has three species Y1, Y2 and Y3. Furthermore, we also know that these two genera share some characters and are therefore considered to be closely related. Based on these pieces of information from traditional taxonomy, we can conclude that species X1, X2 and X3 should ideally form one clade that would be sister to another clade consisting of species Y1, Y2 and Y3 (Fig. 9.3a). We can further refine this tree by identifying other morphological characters that help us resolve the relationships within these two clades (Fig. 9.3b). Figure 9.3b represents our morphology-based hypothesis that depicts the relationships among these six species. We can then

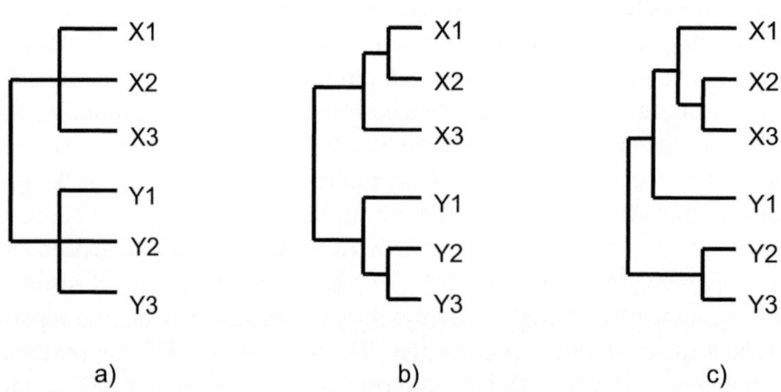

Fig. 9.3 Morphological phylogenetic tree (**a** and **b**) vs. molecular phylogenetic tree (**c**)

generate molecular data for these species and use it to build a molecular phylogeny (Fig. 9.3c).

In the molecular phylogeny, genus X is monophyletic as expected from the taxonomy, but the relationships between species differ when compared to Fig. 9.3b. However, genus Y is not monophyletic in that Y1 is sister to X. Nevertheless, Y2 and Y3 are sister to each other. Thus, the morphology-based tree is similar to the molecular tree in some aspects, but these trees are not identical. This raises the question: Is the phylogenetic tree based on molecular data significantly different from the morphology-based tree?

To address this issue, we calculate the likelihood of the morphology tree, given the molecular data, and compare it with the likelihood of the molecular tree (Fig. 9.4). The molecular data used to generate the molecular tree are also used to calculate the likelihood of the morphological tree. The likelihood of the molecular tree will always be higher than the morphology-based tree, but we are interested to know if these scores are significantly different.

An appropriate test for this comparison is the Shimodaira-Hasegawa (SH) test (Shimodaira and Hasegawa 1999). The SH test calculates the probability that the difference in likelihood scores of the two trees is due to chance. If the difference in the likelihoods of the two trees is significant ($P < 0.05$), then the molecular tree has significantly higher likelihood score than the morphology tree. In other words, the molecular tree fits the molecular data significantly better than the morphology tree. For a detailed treatment of various likelihood-based statistical tests of competing tree topologies, *see* Goldman et al. (2000) and Schmidt (2009). Another test for topological comparison is the AU test (see Shimodaira (2002) for details). These tests can also be used to study character evolution and island biogeography. For example, one might be interested to know if a tree suggesting single origin of a character is significantly better than a tree supporting multiple origins. Similarly,

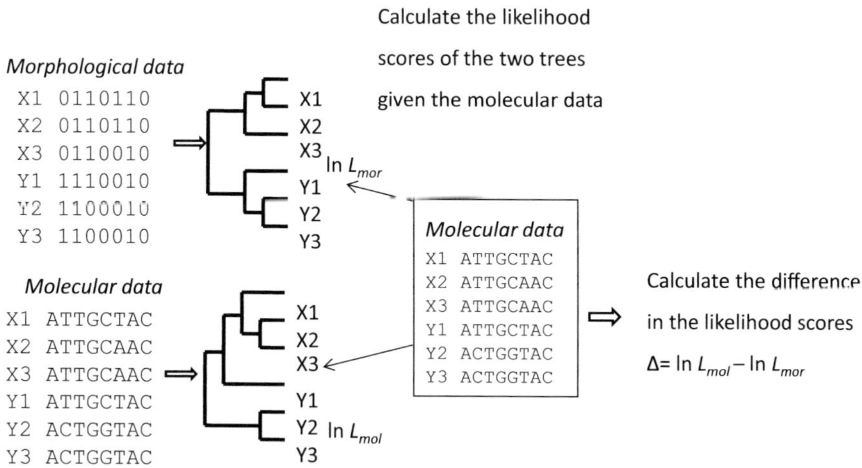

Fig. 9.4 Comparing the likelihoods of two competing trees, given the molecular data

one can use these tests to choose the hypothesis that best explains the origin of island biota, particularly to understand if the island's biota has been assembled through single or multiple colonisation events.

Model Selection

One of the requirements of likelihood-based tree-building methods is model selection. The model of sequence evolution needs to be specified a priori before we can search for the optimal tree. Thus, as a researcher, one has to decide which model best fits the dataset at hand. Models provide us with a set of assumptions regarding the process of nucleotide evolution (Posada 2009). There is a plethora of models that can be evaluated. For example, the programme MODELTEST has a list of 56 models (Posada and Crandall 1998). Chapter 4 lists some of these models. One of the widely used methods for model selection is the likelihood ratio test, which compares the likelihoods of multiple models two at a time and helps pick the model which has a significantly higher likelihood score. To do this, the programme MODELTEST first builds a neighbour-joining (NJ) tree for the data at hand using *p*-distance. The likelihood of this NJ tree is calculated using the simplest model, JC 69 (*see* Chap 4), and the sequence data. The likelihood of the NJ tree is then recalculated using the next more complex model, for example, F81. The difference in the likelihood scores between the parameter-rich model (F81) and the simple model (JC69) is calculated. If this difference is significant, then the complex model is selected over the simpler model. In the next iteration of this programme, F81 becomes our simple model and is now compared with the next more complex parameter-rich model (e.g. HKY 85, *see* Chap. 4). In every step of this search, more and more parameter-rich models are selected. This is because a parameter-rich model fits the data better and therefore will give us a higher likelihood score. However, the number of parameter values to be estimated from the same amount of sequence data also goes up. Thus, the error associated with the parameter estimates becomes an issue. The programme stops when the difference in the likelihood scores between the more complex and simple models is not significant. At this point it selects the simpler model because the more parameter-rich model does not fit the data significantly better than the simpler model. The likelihood ratio test compares two models at a time, whereas other methods for model selection such as Akaike Information Criteria (AIC) and Bayesian Information Criteria (BIC) simultaneously compare all competing models (Posada 2009). With multiple genes increasingly being used for phylogenetic analyses, more complex models that partition genes and codons are being used. One such programme is *PartitionFinder* (see Lanfear et al. (2016) for more information).

References

Goldman N, Anderson JP, Rodrigo AG (2000) Likelihood-based tests of topologies in phylogenetics. Syst Biol 49(4):652–670

Lanfear R, Frandsen PB, Wright AM, Senfeld T, Calcott B (2016) PartitionFinder 2: new methods for selecting partitioned models of evolution for molecular and morphological phylogenetic analyses. Mol Biol Evol 34(3):772–773

Posada D (2009) Selecting models of evolution. In: Lemey P, Salemi M, Vandamme A-M (eds) The phylogenetic handbook: a practical approach to phylogenetic analysis and hypothesis testing, 2nd edn. Cambridge University Press

Posada D, Crandall KA (1998) MODELTEST: testing the model of DNA substitution. Bioinformatics Appl Note 14(9):817–818

Schmidt HA (2009) Testing tree topologies. In: Lemey P, Salemi M, Vandamme A-M (eds) The phylogenetic handbook: a practical approach to phylogenetic analysis and hypothesis testing, 2nd edn. Cambridge University Press

Shimodaira H (2002) An approximately unbiased test of phylogenetic tree selection. Syst Biol 51(3):492–508

Shimodaira H, Hasegawa M (1999) Multiple comparisons of log-likelihoods with applications to phylogenetic inference. Mol Biol Evol 16(8):1114–1116

Applications of Molecular Phylogenetics 10

Abstract

This chapter surveys the wide-ranging applications of phylogenetics. It demonstrates how trees aid in taxonomy, systematics, conservation biology, historical biogeography, phylogeography, epidemiology and the study of character evolution. Examples include detecting cryptic species, tracing dispersal routes, resolving speciation modes and reconstructing the evolution of genes, traits and behaviours. The chapter shows how phylogenies inform conservation priorities by revealing evolutionary distinctiveness and uncovering hidden diversity. The chapter ends with a brief explanation on phylogenetically independent comparison and phylogenetic comparative methods and their applications.

Keywords

Biogeography · Phylogenetic diversity · Vicariance · Dispersal · Phylogenetic comparative methods

Molecular Systematics and Species Delimitation

Perhaps one of the most important and extensive uses of molecular phylogeny is in taxonomy. With more and more phylogenies being generated for different taxonomic groups, researchers are realising that molecular phylogenies are often not congruent with morphology-based classifications (taxonomy). This incongruence between morphology and molecular data has been reported at various levels of classification. In some cases, the incongruence is at the species level, i.e. species assigned to a genus often do not fall in the same clade in the phylogeny, or it could be at higher levels. In such cases, the taxonomy is revised, such that it is consistent with the phylogeny. The term 'molecular systematics' is often used to describe this endeavour. Take the example of the taxonomy of langurs of South Asia. Traditionally, at least six species of langurs have been described from South Asia. These include the langurs found in the Indian subcontinent—Hanuman langur, Nilgiri langur and purple-faced langur; the langurs found in the northeast states of India, Bhutan and

K. P. Karanth, *Molecular Phylogenetics*, https://doi.org/10.1007/978-981-95-9147-3_10

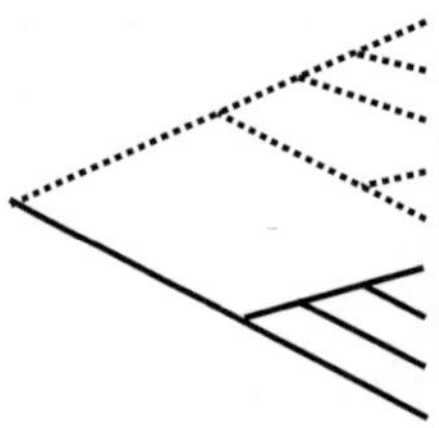

	Morphology	Molecular data
Hanuman langur	*Semnopithecus*	*Semnopithecus*
Nilgiri langur	*Trachypithecus*	*Semnopithecus*
Purple-faced langur	*Trachypithecus*	*Semnopithecus*
Capped leaf monkey	*Trachypithecus*	*Trachypithecus*
Golden leaf monkey	*Trachypithecus*	*Trachypithecus*
Phayre's leaf monkey	*Trachypithecus*	*Trachypithecus*
Dusky leaf monkey	*Trachypithecus*	*Trachypithecus*
Francois' leaf monkey	*Trachypithecus*	*Trachypithecus*
Silvered leaf monkey	*Trachypithecus*	*Trachypithecus*

Fig. 10.1 Molecular systematics of the langurs and leaf monkeys of Asia. The phylogeny is based on nuclear and mitochondrial markers. Also shown are the morphology-based classification and the revised taxonomy based on molecular data. NI: North India; SI: South India; SL: Sri Lanka

Bangladesh (northeast)—golden langur and capped langur; and Phayre's leaf monkey that is distributed predominantly in mainland Southeast (SE) Asia, but its distribution also extends to the state of Tripura in the northeast and parts of Bangladesh (Karanth et al. 2008). Most classification schemes place the Hanuman langur in the monotypic genus *Semnopithecus* and the remaining species in the genus *Trachypithecus,* along with the leaf monkeys of SE Asia (Fig. 10.1). However, molecular data do not support this arrangement. On the contrary, a range of nuclear and mitochondrial markers suggest that the langurs of the Indian subcontinent, i.e. Hanuman langur, Nilgiri langur and purple-faced langur, are closely related. Accordingly, in the revised taxonomy, based on molecular data, Nilgiri langur and purple-faced langur are also placed in the genus *Semnopithecus* along with Hanuman langur. The remaining langurs are placed in genus *Trachypithecus.*

Molecular phylogenetic studies have also helped us resolve another problematic issue in taxonomy, namely, identification and delimitation of closely related and morphologically similar species (Karanth 2017). For example, the Hanuman langur, a widely distributed taxon, was long considered a single species by most taxonomists. However, molecular studies suggest that this taxon constitutes a complex with multiple species (Fig. 10.1). Often, molecular studies have been the first line of evidence, suggesting the presence of cryptic species, i.e. taxa that exhibit little or no discernible morphological difference. Once these putative taxa are identified, additional lines of evidence, such as ecological data and behavioural data, are used to delimit species. This approach, which espouses the use of multiple lines of evidence to delimit species, is called integrative taxonomy and has been instrumental in characterisation of cryptic diversity (see Karanth 2017 for details). Thus, molecular phylogenies have been a very important tool for species delimitation and by extension for better understanding of the process of speciation. For example, knowing how species are related is very important to determine the mode of speciation (allopatric, peripatric, parapatric and sympatric).

Another interesting application of phylogenetics, in taxonomy, has been the proposal to replace traditional rank-based nomenclature with phylogenetic nomenclature (de Queiroz and Gauthier 1994). The authors point out that the non-evolutionary basis of the current taxonomic system often causes ambiguity and instability in nomenclature. This scenario could arise when there is a taxonomic revision or when there is difference of opinion among taxonomists. Phylogenetic nomenclature proposes a rank-free clade-based naming of various taxonomic groups. In this system of nomenclature, taxa names are explicitly defined based on common ancestry (see de Queiroz 2012 for more details). The rules governing phylogenetic nomenclature are given in *phylocode* (https://www.ohio.edu/phylocode/).

Phylogenetic Diversity

One of the most commonly used measures of biodiversity is 'species richness'. *Species richness* (SR) tells us how many different species there are in an area, and areas with higher SR are often accorded higher priority for conservation. However, SR does not take into consideration the evolutionary history encompassed in an area. For example, an area with four species of lizards (skink, gecko, chameleon and garden lizard) and another area with four species of reptiles (gecko, cobra, crocodile and tortoise) will both be assigned the same value for reptile richness. Nevertheless, the latter has greater diversity as it harbours a very diverse composition of reptiles. Areas harbouring more distantly related species encompass a greater evolutionary history and therefore a higher biodiversity. This information is not incorporated in SR, and therefore traditional measures of diversity often underestimate true diversity. In this regard, phylogenetic diversity (PD) that takes into consideration evolutionary histories of species in an area, by incorporating information regarding their relationships in calculating diversity, has gained much traction in the last decade. In the broad sense, PD is the sum of branch lengths in a phylogenetic tree of species from a given area (see Karanth et al. 2019). Numerous studies have noted that SR and PD are often decoupled and that SR cannot be used as a surrogate for PD (Divya and Karanth 2018). In recent years, PD has been extensively used in conservation biology, particularly for prioritising areas for conservation. Furthermore, PD has also been used to understand broad-scale historical patterns, such as in biogeography of a region as well as local community assemblages. A robust and well-supported phylogeny is central for accurate calculation of PD. For a more detailed treatment of this topic, see Karanth et al. (2019) and Divya and Karanth (2018) (Fig. 10.2).

Historical Biogeography

Biogeography is the study of geographical distribution of biodiversity, i.e. what are the patterns of species distributions across large geographical areas. Furthermore, biogeographers are interested in understanding the historical processes (past events) that have generated these patterns. Current distributions of extant taxa are determined by both intrinsic factors such as species biology, life history traits and extrinsic factors such as the presence of competitors and suitable habitats. Additionally, species distributions have also been shaped by events in the past. It is these past

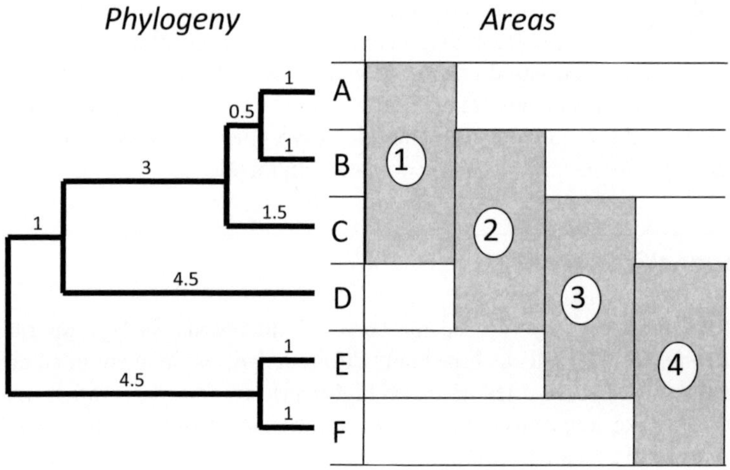

Fig. 10.2 A hypothetical phylogeny showing the relationships between species A and F. To the right of the phylogeny are the areas (1 to 4) where these species are distributed. For example, area 1 harbours species A, B and C. Numbers on the phylogeny represent changes along each lineage. All areas have the same SR (three species); however, their PDs vary. PD is the sum of branch lengths (SBL) from the root in a phylogeny of species from an area, for example, SBL of species A, B and C from *area 1* is 8, whereas for *area 3*, it is 15.5 (see Divya and Karanth 2018; Karanth et al. 2019)

events that biogeography attempts to decipher. Traditionally, biogeography looked at distribution patterns across multiple species and assigned biota into various biogeographical regions and sub-regions. With the advent of phylogenies, a whole new window has opened up in the study of biogeography. The use of phylogenetic tools in biogeography is termed '*historical biogeography*'.

The two broad and often contrasted processes invoked to explain spatial and temporal patterns of biotic diversity are vicariance and dispersal (Zink et al. 2000). According to the 'Vicariance school', species arise when a widely distributed ancestral species gets isolated into smaller populations due to the emergence of a barrier between them. These barriers (vicariant events) could be the uplifting of mountain ranges, continental drift, climate change, etc. Whereas the 'dispersal school' postulates that a subpopulation of an ancestral species disperses across a pre-existing barrier, becomes isolated and eventually evolves into a different species (see Fig. 10.3). The fundamental difference between these two scenarios is that vicariance speciation is concomitant with the emergence of the barrier, whereas the emergence of the barrier predates speciation in dispersal speciation.

The Indian subcontinent is an interesting landmass to understand the role of vicariance and dispersal in shaping the evolution of its biota. This is because much of this landmass, along with Africa and Madagascar, was part of the Gondwana supercontinent, around 200 million years ago (mya). The Indo-Malagasy plate separated from Africa around 110 mya, followed by the split between India and Madagascar around 80 mya. Given this geological history, one plausible scenario is

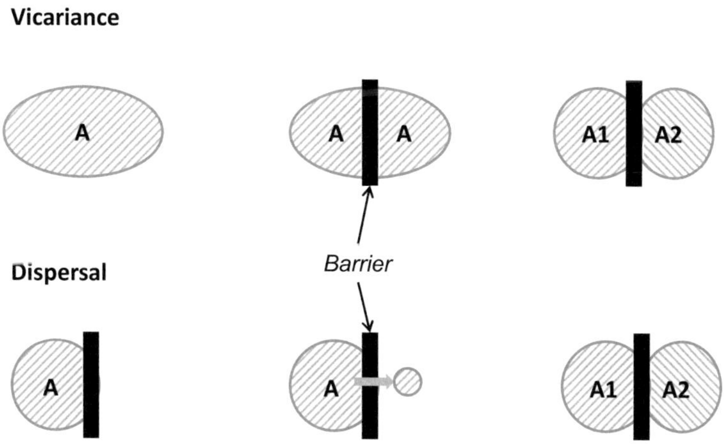

Fig. 10.3 Vicariance vs. dispersal: the letter A represents ancestral species that diverged into daughter species A1 and A2 on either side of the barrier. In case of vicariance, the speciation is concomitant with the emergence of the barrier, whereas in case of dispersal, the barrier predates the speciation event

that a taxon, widely distributed in Gondwanaland, got isolated in Africa, Madagascar and India, when this supercontinent broke up. This would be the classic vicariance scenario, and the resulting topology of the taxa would mirror the sequence and time of separation of these Gondwanan fragments (Fig. 10.4a). Alternatively, species could have dispersed into India much after the formation of the oceanic barriers (110 and 80 mya). These dispersals could either be transoceanic from Madagascar/Africa (Fig. 10.4b1–b2) during the journey of the Indian plate towards Asia or due to geodispersal from Africa via Arabia (Fig. 10.4b3).

These scenarios can be explicitly tested using dated molecular phylogenies (see Datta-Roy and Karanth 2009). Thus, phylogenies are critical in determining how the biota of India (or any region) was assembled. In the Indian context, it has helped us better understand the origin of Indian biota, i.e. where did they come from? When and how (geodispersal or transoceanic dispersal)? Furthermore, phylogenies are also useful to explore the biogeography of taxa after they got here. In other words, what are the more past palaeoclimatic and palaeogeological events that might have shaped the current distribution of taxa in India? (see Karanth 2003, 2021 for details).

Other Applications of Molecular Phylogenetics

There are numerous other applications of phylogenetics, and I hope to discuss these applications in the next edition. I will briefly mention some of these applications. One of the most important and interesting applications of phylogenetic trees is in understanding character evolution. The idea is to reconstruct the ancestral character

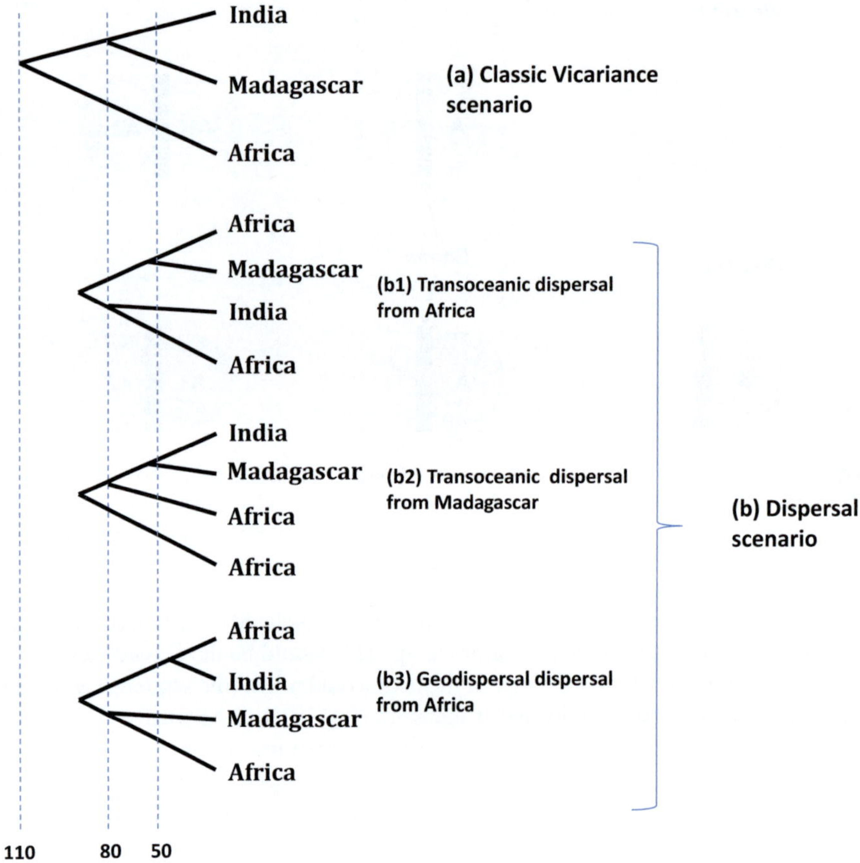

Fig. 10.4 Phylogenetic predictions under vicariance and dispersal scenarios. The numbers 110 and 80 refer to the dates in million years ago (mya) and correspond to the time of separation Africa-Indo-Madagascar and India-Madagascar, respectively. The merging of the Indian Plate with Asia was around 50 mya, which facilitated geodispersal from Asia (and Africa)

states for the internal nodes in a phylogenetic tree and use this information to understand where, when, how and why certain states evolved along certain lineages. The character in question can be morphological, behavioural, molecular and developmental. Thus, this kind of analysis has applications in the fields of morphological and behavioural evolution, molecular evolution and evolutionary developmental biology. Phylogenies are indispensable in studies pertaining to co-speciation. Co-speciation is a process wherein the host phylogeny closely mirrors the phylogeny of its obligate symbiont/pathogen/parasite. Discordance between these phylogenies is indicative of a host shift. Such studies have been extensively used to understand the origin and evolution of zoonotic diseases. Today, phylogenies are being extensively used in statistical analysis for phylogenetically independent comparison (PIC). Just as data points close to each other in space tend to be similar (spatial autocorrelation), related species tend to have similar trait values and therefore are not independent of each other. Thus, phylogenetic information is needed to control for similarity due to evolutionary relationships in a statistical analysis. Lastly, phylogenies are being increasingly used to study macroevolutionary processes, an area referred to as phylogenetic comparative methods (PCM). In PCM, one is interested in studying the rate of diversification within and across clades, i.e. has the rate of diversification been uniform, increasing or decreasing over time and what are the factors that might be driving these changes (climate, novel trait, etc.). See Cornwell and Nakagawa (2017) for details.

References

Cornwell W, Nakagawa S (2017) Phylogenetic comparative methods. Curr Biol 27:R327–R338

Datta-Roy A, Karanth KP (2009) The out-of-India hypothesis: what do molecules suggest? J Biosci 34(5):687–697

de Queiroz K (2012) Biological nomenclature from Linnaeus to the PhyloCode. In: Bell CJ (ed) The herpetological legacy of Linnaeus: a Celebration of the Linnaean tercentenary, bibliotheca Herpetologica, pp 135–145

de Queiroz K, Gauthier J (1994) Towards a phylogenetic system of biological nomenclature. Trends Ecol Evol 9(1):27–31

Divya B, Karanth KP (2018) Capturing richness independent phylogenetic diversity and testing surrogates for PD in woody plant communities. Curr Sci 115(5):910–919

Karanth KP (2003) Evolution of disjunct distribution among wet zone species of the Indian subcontinent: testing various hypothesis using a phylogenetic approach. Curr Sci 85(9):101–108

Karanth KP (2017) Species complex, species concepts and characterization of cryptic diversity: vignettes from Indian systems. Curr Sci 112(7):1320–1324

Karanth KP (2021) Dispersal vs. vicariance: the origin of India's extant tetrapod fauna. Front Biogeogr. https://doi.org/10.21425/F5FBG48678

Karanth KP, Singh L, Collura RV, Stewart C-B (2008) Molecular phylogeny and biogeography of langurs and leaf monkeys of South Asia (primates: Colobinae). Mol Phylogenet Evol 46:683–694

Karanth KP, Gautam S, Arekar K, Divya B (2019) Phylogenetic diversity as a measure of biodiversity: pros and cons. J Bombay Nat Hist Soc 116:53

Zink RM, Blackwell-Rago RC, Ronquist F (2000) The shifting roles of dispersal and vicariance in biogeography. Proc R Soc Lond B Biol Sci 267:497–503

11 Molecular Phylogenetics: Frequently Asked Questions

Abstract

The final chapter addresses common questions that crop up while initiating and executing molecular phylogenetic studies. Topics include marker choice, sampling strategy, which tree-building method to use, treatment of gaps, etc. Further, the chapter explores interpretation of trees and bootstrap values, causes of tree discordance and practical concerns in data analysis. It synthesises earlier material into accessible answers for students and researchers while pointing to pitfalls and best practices. The chapter stresses the importance of using multiple markers, appropriate models and careful interpretation. By anticipating user challenges, it provides practical solutions to typical issues one faces in phylogenetic studies.

Keywords

Sampling · Discordance · Gaps · Missing data · Molecular markers

How do I start?

Before embarking on a phylogenetic study, ask yourself, 'what is the question I am trying to address?' How will phylogeny help me address this question? If you are not able to come up with clear-cut answers to these questions, then your study design itself might be flawed. Let us say that you are interested in resolving the taxonomy of species in a genus. There are many taxonomic schemes that have been proposed wherein the species in this genus have been assigned to different species groups. You want to find out if the phylogeny is concordant with these species groups.

A practical approach is to start with sequencing one mitochondrial (for animals) or chloroplast (for plants) gene from one specimen per species. These are the most widely sequenced markers, and primers are easily available. Try to sample as many (if not all) species in a genus. Ideally, one should cover around 70% of the species if one is looking at species-level relationships. Remember that some or a significant

K. P. Karanth, *Molecular Phylogenetics*, https://doi.org/10.1007/978-981-95-9147-3_11

number of species in your genus of interest might be distributed outside India. These must also be included; otherwise you will run the risk of incomplete taxon sampling (*see* 'How many species to sample?' section below). Phylogenies are not confined by political boundaries. There are two ways to resolve this issue: either choose groups for which sequences are available from other parts of the world or collaborate with researchers abroad. Make sure to include an appropriate outgroup in the sampling plan. The choice of the marker depends on the question being asked (see below). In our example, we are looking at closely related species. Therefore, start with mitochondrial/chloroplast genes. Look for mitochondrial/chloroplast gene sequences in GenBank for your genus or closely related genera and pick the gene for which a large number of sequences across taxonomic groups is available. Sequence this gene from your taxonomic group. This should give you a rough picture of the overall phylogenetic pattern in the species group of interest. Depending on the availability of resources and time, the project could be expanded to include at least two specimens per species and for widely distributed species multiple specimens (around 5) from across the range. Once species monophyly is established, multiple nuclear markers can be included. This step is necessary in cases where species delimitation is problematic or species identification is not unambiguous. In the case of nuclear markers, one exemplar specimen from each species or monophyletic group should be adequate. If the individual phylogenies derived from different markers are largely congruent with each other, these individual datasets can be combined, and a single tree can be generated from the combined dataset. If in this phylogeny, certain nodes have low bootstrap support, then additional markers could be included. These issues are discussed in detail below.

How many markers to use?

One can use a single marker or multiple markers, all the way to the complete genomes to derive a phylogeny. A phylogeny based on a single marker is often not a good estimate of the species phylogeny because in some cases a particular gene tree might not correspond to the species tree (*see* Chap. 8, Gene trees vs. species trees). Each gene tree tells us the evolutionary history of that gene in the species group from which it has been sequenced. This tree might not be congruent with the species tree, in that some parts of the tree might be different. If the chosen marker happens to fall in this category, then our estimate of species phylogeny will be erroneous. We won't know this until additional markers are sequenced. Moreover, a single marker covers a very small part of the genome and does not represent the evolutionary pattern/s of the rest of the genome. Thus, multiple markers should be used so that one can either present a consensus phylogeny of multiple phylogenies based on different markers or a single phylogeny based on the combined dataset. In the ideal world, complete genomes should be sequenced. This option is currently not feasible to most researchers, as it is prohibitively expensive to sequence complete genomes. Additionally, handling as well as analysing such large datasets is computationally difficult. Therefore, we have to, due to practical reasons, stay with the multiple markers approach. This brings us back to our question as to how many markers to use. There are two ways to ask this question.

Firstly, how many markers should one use to estimate the 'true' phylogeny?

There is no way of knowing the 'true phylogeny' unless we go back in time and record the various speciation events in the species group of interest. Let us assume for the sake of argument that we did know the true phylogeny. If so, we can then keep adding markers to our phylogenetic analysis till we get a tree that is identical to the true phylogeny. Under this unrealistic condition, we can address the above question. But typically, we do not have the true phylogeny. Instead, we are attempting to estimate species phylogeny based on our knowledge of how genes evolve. Nevertheless, there are statistical tools that give us a measure of the robustness of our estimates. Thus, to address this issue, we take the second approach by asking the question:

How many markers should one use to get a 'robust' or well-supported estimate of a phylogeny?

This question can be addressed by implementing bootstrap search. Typically, a well-supported phylogeny has bootstrap support ranging from 70% to 100% for various nodes, the assumption being that a well-supported phylogeny is a good estimate of the true phylogeny. If the bootstrap support is less than 70% for certain nodes, then additional markers or more sequence data from a given marker need to be added to the dataset. Thus, bootstrap support could be used as an indicator to determine the upper limit for the number of markers that one needs to use to derive a robust phylogeny. Recent works have indicated that for most studies, usually ten nuclear markers should suffice. The point to note here is that the decision on what kind of markers to use will depend on the position of the nodes with low bootstrap support. If the node is situated close to the root (higher taxonomic level), then a slowly evolving marker should be used. On the other hand, if the node appears closer to the terminal nodes (tip of the evolutionary tree), then a rapidly evolving marker should be used (*see* section on What kind of markers to use?). Another point to bear in mind is that in situations where there is short internodal time or deep coalescence (*see* Chap. 8), the bootstrap support will always be low, and adding more data might not resolve the issue. Nevertheless, the decision on the number of markers to use is often driven by availability of time and resources.

How do we know that the tree-building methods are generating the right phylogeny? Accuracy of phylogenetic reconstructions

Assuming that the chosen gene is a perfect marker for phylogenetic reconstruction, how can we be sure that the various tree-building methods are reconstructing the species phylogeny accurately. In the previous question, we assume that tree-building methods are accurate in reconstructing evolutionary histories and we wanted to determine the quantity of molecular data needed to derive a robust phylogeny. On the other hand, here we assume that adequate molecular data are available, but we need to know whether the tree-building methods are retrieving the right phylogeny. Here again, knowledge about 'true phylogeny' of a species group would be useful. The true phylogeny can be compared with the phylogenies generated by various tree-building methods to determine the accuracy of their reconstructions. However, as mentioned above, it is impossible to know the true phylogeny unless we have the ability to go back in time. So how do we address this issue? Researchers have addressed this issue using both simulated data and experimental data (*see* Hillis 1995 for other approaches). In simulation studies, sequence data are simulated for a given tree and model of sequence evolution. Various tree-building methods are used to estimate trees from these simulated data. These trees are compared with the given topology to determine phylogenetic accuracy. Simulated data have two issues: firstly, the result is biased towards the model of evolution used, i.e. methods that perform better for a particular model will generate accurate phylogenies, while others will not perform as well. Secondly, simulated data do not incorporate all the complexities of a biological system. In the experimental approach, laboratory lineages of viruses are experimentally generated such that the true phylogeny is known. These experimentally generated lineages are sequenced and the sequences are then used to test the accuracy of various methods. The advantage of experimental approach is that it does not have some of the issues of simulated data. These studies have shown that most tree-building methods do retrieve the right tree. (*see* Chap. 12 of Molecular Systematic by Hillis et al. 1996 for more information).

What kind of markers to use?

Choice of markers depends on the question at hand. If we are interested in species-level phylogeny (closely related and recently diverged species), then one needs to use rapidly evolving markers such as mitochondrial genes in the case of animals. On the other hand, for higher-level phylogenies (genera and above), one needs to use slowly evolving markers such as nuclear markers. In mammalian systems, mitochondrial markers typically evolve five to ten times faster than nuclear markers, whereas in plants, mitochondrial markers evolve slower than nuclear markers. The extranuclear markers of choice in plants are chloroplast markers.

Neutral markers ersuvs markers Under selection

Ideally, neutral markers (non-coding regions) should be used, but these are difficult to find. Most studies use protein-coding genes as a huge number of coding regions have been characterised for hundreds of species. However, coding regions are, by definition, under selection! Nevertheless, using protein-coding genes in phylogenetic analysis is not necessarily problematic. For example, if the gene is under negative (purifying selection), then there is no serious issue. This is because in a gene under negative selection, any nonsynonymous change is purged, and only silent changes remain. Silent/synonymous changes, by definition, are neutral (usually changes in the third position of the codon). One can ascertain where all the changes are occurring in a dataset by using programmes like MEGA. If most of the phylogenetically informative changes are synonymous changes, then our phylogenetic inference is largely based on neutral changes (which are what we want). If the gene is under positive selection, then it could be a problem due to issues such as convergence, long branch attraction and accumulation of apomorphies. Under such a situation, nonsynonymous changes should be dropped from the analysis.

Sex chromosome markers versus Autosomal markers versus mitochondrial markers

Mitochondrial and Y-chromosomal markers are haploid and undergo uniparental inheritance. Therefore, their effective population size is one fourth that of nuclear autosomal genes (*see* Chap. 7). This means they attain reciprocal monophyly much faster than autosomal markers. This attribute of these markers is useful for resolving relationships between closely related and recently diverged species (*see* Moore 1995 for details). Nevertheless, these markers give us only part of the story, i.e. male or female lineage of a species. These markers also constitute a very small part of the genome. Autosomal markers are spread over the entire genome and undergo biparental inheritance. Thus, different classes of markers have different evolutionary histories due to differences in the mode of inheritance, effective population size, different selection pressure, mutation rate, etc. Therefore, multiple kinds of markers should be used to get the complete picture.

The rule of thumb I use for the mammalian system is one Y/X chromosome marker, 4–5 autosomal markers and one mtDNA marker. In the case of plants, one can use couple of chloroplast markers and four to five nuclear markers. Often, marker choice is driven by what is available in GenBank. The chosen marker(s) should have adequate variation to address the question at hand. As mentioned above, bootstrap support can be implemented to determine whether the variation is adequate.

In the case of multiple markers, do we combine the different datasets or analyse them separately?

To start with, each marker should be analysed separately to determine whether trees generated by different datasets are congruent with each other or not. Additionally, partition homogeneity test (also called incongruent length difference) can also be implemented to test whether the topology of the trees generated by different datasets are significantly different from each other. If the above exercise suggests that trees generated by different datasets have similar overall tree topology and the difference between them is not significant, then the datasets can be combined. If not, these datasets should be analysed separately (*see* below for more information). However, if the incongruence is due to deep coalescence (Chap. 8), then use species-tree approach (see Chap. 6).

What do we do if different markers generate different tree topologies?

At the onset, it must be noted that incongruence between datasets (e.g. between a mitochondrial and a nuclear gene) does not mean that the tree-building methods

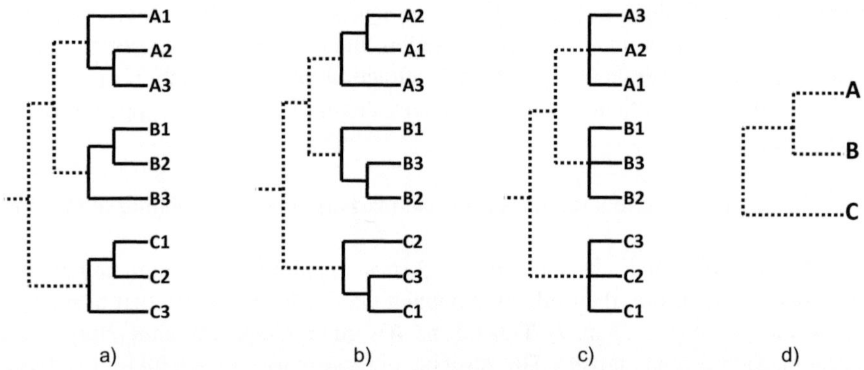

Fig. 11.1 Incongruence between two genetic markers. (**a**) and (**b**) show phylogenetic trees built using two different markers. (**c**) shows the phylogenetic tree using the combined dataset. Dotted lines indicate the same generic-level relationships retrieved by each tree, shown in (**d**)

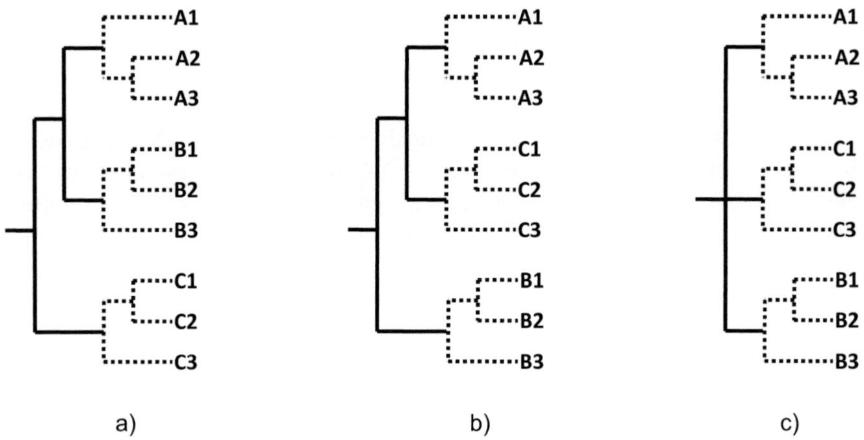

Fig. 11.2 Incongruence between two genetic markers. (**a**) and (**b**) show phylogenetic trees built using two different markers. (**c**) shows the phylogenetic tree using the combined dataset. Dotted lines indicate the same species-level relationships retrieved by each tree

have generated an incorrect tree or that there is a problem with one of the datasets. Whether such incongruence between datasets is a problem or not depends on the question we are trying to address. For example, Figs. 11.1 and 11.2 represent two phylogenetic trees generated using two different markers for the same nine species belonging to three different genera A, B and C. In Fig. 11.1a and b, the relationships between genera are identical (dotted lines), whereas species relationships within each genus are different. If we are interested only in relationships between genera, and relationships among species within each genus do not concern us, then we can combine the datasets. The combined dataset provides the broader, higher-level phylogeny (Fig. 11.1c) that shows the generic-level relationships (Fig. 11.1d).

Alternatively, one might be interested in relationships within a genus, and between-genera relationships might not be of concern (Fig. 11.2). In the example shown below (Fig. 11.2a, b), species relationships within each genus (A, B, C) are identical in the two trees, but relationships between genera are different. Here again, datasets can be combined if within-genus relationships are identical across datasets (Fig. 11.2c).

Thus, it depends on where in the tree the difference has appeared and to what extent this incongruence will affect your hypothesis. Nevertheless, there might be cases where multiple markers provide us with very different phylogenetic positions of one or more species. If the species in question is of interest to us, then what do we do? Incongruence across datasets is usually indicative of a complex evolutionary history of the species in question. This incongruence might be due to a variety of reasons including past or current hybridization (*see* Karanth et al. 2008), horizontal gene transfer, incomplete lineage sorting or retention of ancestral alleles (*see* Chap. 7). For example, the phylogenetic position of species 'X' in Fig. 11.3a, b is very different. We can either show its position as unresolved (Fig. 11.3c) or if we suspect hybrid origin, we can show it as being equidistant from A and B. These scenarios should be explored further with additional markers. Furthermore, other kinds of data, such as ecological, behavioural and morphological data should be utilised to clarify this issue. Trees generated from different markers can be either presented separately or a consensus tree can be generated.

Which tree-building method to use?

Each tree-building method makes a different set of assumptions and has its strengths and weaknesses (*see* section on tree-building methods). Additionally, no single method is better than the rest, and each method uses a different approach to reconstruct evolutionary history. Thus, the best strategy is to use as many methods as possible, particularly methods that use optimality criteria. Fortunately, in most cases, different tree-building methods tend to give similar results. When multiple

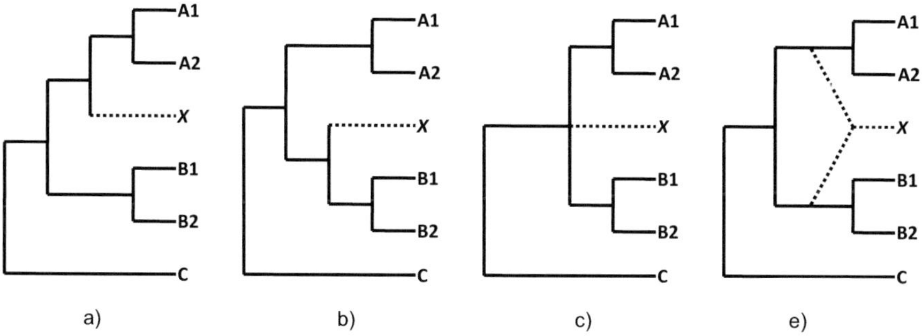

Fig. 11.3 Different phylogenetic positions of species X, retrieved in different trees based on different markers (**a** and **b**). (**c**) depicts the position of species X as unresolved. (**d**) depicts species X as a hybrid of species A and B

tree-building methods give congruent results, we have greater confidence in our phylogeny. Generally, phylogenetic studies use at least two different methods—maximum likelihood or Bayesian approach. Of late, parsimony has gone out of fashion as it does not use a model of sequence evolution. Unweighted pair group method with arithmetic mean (UPGMA) and NJ methods are rarely used as these methods do not use an optimality criterion.

What do we do if different tree-building methods generate different tree topologies?

Ideally, you should report the different results of both the trees. The question is, does this incongruence between methods have any bearing on the question you are trying to address? If so, then you have a problem. If not, report both trees and discuss the common elements of the two trees. If resolving the higher-level relationships is important for the question that you are addressing, then ideally you should generate more data from slowly evolving markers. If this is not possible, then you report both trees and mention that one of the trees supports the hypothesis while the other does not and that for a more rigorous test of this hypothesis one has to add more sequence data.

How many samples per species?

One of the fundamental assumptions of molecular phylogenetics is that interspecific (between species) genetic distance is far greater than intraspecific (within species) genetic distance for a given marker. This is because members of a species by definition are genetically (and morphologically) more similar to each other than they are to members of a sister species. For example, in the case of butterflies, it has been found that within-species genetic distance for mitochondrial cytochrome oxidase I (CO-I) gene is less than 2%, whereas between species, it is 10–20%. Thus, between species sequence difference can be five to ten times higher than within species sequence difference. For nuclear genes, this contrast in inter- and intraspecific variation is not as dramatic as for mitochondrial markers. Nevertheless, intraspecific variation is significantly higher than interspecific variation. How do these inter- and intraspecific variations influence the phylogeny? In a phylogenetic tree, individuals/samples of the same species will cluster together because they are genetically similar to each other (low intraspecific variation), and the relationships between these clusters, which is based on interspecific variation, would provide us with the species phylogeny (Fig. 11.4a). However, even when only one sample/species is used, one would get the same tree topology. Therefore, a phylogenetic study where multiple samples/species are used does not provide us with additional information, nor does it generate a different phylogeny when compared with a phylogenetic study where single sample/species are used. Thus, only one sample per species should suffice.

To better understand how inter- and intraspecific variations might influence phylogenetic reconstruction let's take the example of the human-ape phylogeny (Fig. 11.4). The tree shown above has long been established with multiple lines of

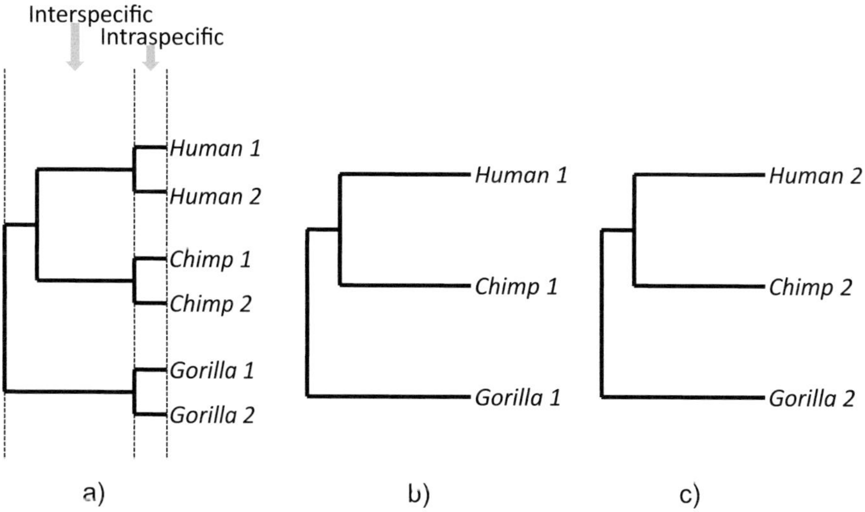

Fig. 11.4 (**a**) Human-ape phylogeny using multiple samples per species. (**b** and **c**) Single sample per species

evidence including molecular data. In this tree, human is most closely related to chimp. In other words, chimps are more closely related to humans than they are to gorillas or any other apes. If we sequence *COI* gene from one human (say a European) and one sample each of the ape species and build phylogenetic tree (Fig. 11.4b, c), we would get the same tree as in the case where multiple samples have been used (Figs. 11.4a). At this point, one might ask the question that if another human sample is used (say an Indian or a Chinese), would we get a different phylogeny? In other words, if a different individual/specimen of a species is used, then by chance alone might we get a different phylogeny? This is a highly unlikely scenario because this would mean that some humans (say Indians) might branch with Gorilla! A scenario like that is possible only if the human specimen used is a 'mutant', i.e. the COI sequence of this individual is very different from other humans. Again, such a situation is highly unlikely because individuals with a high number of mutations in a protein-coding gene will not survive. Thus, even by chance alone, such a situation cannot arise.

However, it must be pointed out that in taxonomically problematic groups, multiple samples/species must be included. In the example discussed above, there is no taxonomic ambiguity with respect to the identity of human, chimp and gorilla; thus, one sample/species will suffice. In situations where taxonomic identity is unclear or where one is dealing with a species complex, it is imperative that we use multiple samples/species. Take the example of langurs of the Indian subcontinent. According to traditional taxonomy, there are three species of langurs in the Indian subcontinent. These include Hanuman, Nilgiri and purple-faced langurs. Among these species, the Hanuman langur is widely distributed in India and is also found in Sri Lanka and inhabits a range of habitats. They also exhibit much morphological variation and have been split into multiple subspecies. Nilgiri and purple-faced langurs, on the other hand, have restricted distribution and are confined to wet forests of Southwest India and Sri Lanka, respectively. In a molecular study done on these langurs (Karanth et al. 2008, 2010), it was found that the so-called Hanuman langurs sampled from across the range were not branching together (were not monophyletic). Instead, Hanuman langurs from Sri Lanka branched with purple-faced langurs also from Sri Lanka and Hanuman langurs from South India branched with Nilgiri langurs, also from South India (Fig. 11.5a). This is in contrast with the human-ape phylogeny, wherein all humans branched together. Today, we know that the so-called Hanuman langur is not a single species. On the contrary, it consists of at least three species—one each from North India, South India and Sri Lanka. Thus, this is a case of species complex consisting of multiple closely related species. If one had used only 'Hanuman langur' from Sri Lanka in the phylogeny of Indian langurs, then one would get a different tree topology (Fig. 11.5b), whereas including 'Hanuman langur' from North or South India would generate a different phylogeny (Fig. 11.5c). Importantly, both of these phylogenies (Fig. 11.5a, b) are very different from the true phylogeny (Fig. 11.5a). Thus, in the case of widely distributed or taxonomically problematic species groups, it is important to include multiple samples/species.

In summary, for most cases, one sample/species should suffice, but if time and resource permit, at least two samples/species could be included in the analysis. These samples should be collected from as far apart from each other as possible, i.e. from the far ends of the species' distribution. In the case of widely distributed or taxonomically problematic groups, include at least five samples/species collected from throughout the range of the species. This is done to incorporate maximum possible intraspecific variation with least possible sampling.

Should we only use type specimens or samples from type locality?

Some taxonomists argue that one should always use type specimens in a phylogenetic study because type specimens are true representatives of a species. Museums

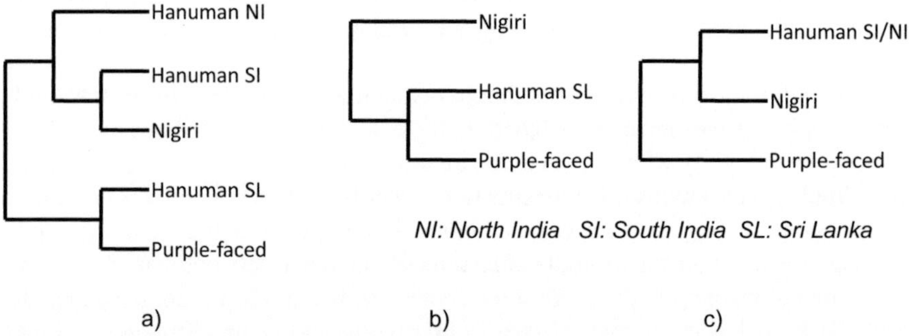

Fig. 11.5 (**a**) True phylogeny for langurs using Hanuman langurs from North India, South India and Sri Lanka. (**b** and **c**) show the different phylogenies generated when Hanuman langurs from only Sri Lanka and only North or South India, respectively, are used

are often reluctant to provide samples of type specimens as the process of sampling might partly destroy the specimen. Moreover, older type specimens have routinely been preserved in formalin, which causes extensive DNA damage. Thus, isolation of high-quality DNA from formalin-preserved type specimens might be problematic. An alternative approach is to collect specimens from the type locality. But if the species has gone extinct from the type location due to anthropogenic activity, as has happened in many cases, this approach cannot be implemented. In such situations, one can attempt to collect the specimen from a location closest to the type locality. Nevertheless, this requirement of including type specimens or type locality specimens in a phylogenetic study is often not practical.

The issue here is about unambiguous identification of previously described species through good taxonomy. Multiple steps must be rigorously employed for the unambiguous assignment of a specimen to a species. These include tentative assignment of a specimen to a species using published keys, followed by consultation with experts working on that taxonomic group and comparison with the type specimen when possible. Taxonomists often argue that the specimens collected outside type locality might not represent the same species as the type specimens. This might be an issue in the case of species complexes that harbour multiple cryptic species (such as Hanuman langur). In such a situation, it is plausible that specimens collected outside type locality might represent a different/new species. This is an issue of incomplete/incorrect taxonomy of a taxonomically problematic group. Here, the first step should be to resolve the taxonomy of the species complex and then proceed with phylogenetic work. Alternatively, phylogenetic tools could be applied to identify these putative cryptic species, i.e. phylogenetics could be used to resolve some of these taxonomic issues. If this is the goal, then it must be clearly stated, and in such studies, including type specimens or type-locality specimens is crucial to address the aforementioned issues.

In species groups for which the taxonomy is well resolved and species identity is unambiguous, there is no need to analyse type specimens or type locality specimens. For example, the type locality for tiger (*Panthera tigris*) is Bengal. If we are interested in estimating the phylogeny of the greater cats which include all the large cats such as lions, leopards, tigers etc., is it necessary to include the tiger sample from Bengal, or will a sample from Bandipur National Park in South India do? We know that the large striped cat from Bandipur is a tiger and not any other species. In fact, we also know that all the tigers in India belong to the same subspecies. The taxonomic identity of this specimen from South India is unambiguous. Thus, in this case, there is no need to include either a type specimen or a type locality specimen in our phylogenetic study. The important point here is that the species identification should be unambiguous.

How many species to sample?

Taxon sampling is one of the fundamental issues in phylogenetics. In the earlier section, the importance of including multiple samples per species in the case of widely distributed or taxonomically problematic species was discussed. Here, we look at how incomplete taxon sampling at the species level can be problematic. Most phylogenetic studies deal with species-level phylogeny wherein one is interested in the evolutionary relationships among species in a genus or among species in a species group. Ideally, in such studies, as many (if not all) species from across the range of the genus must be sampled.

Figure 11.6 illustrates the problem associated with incomplete sampling of species in a phylogenetics study. In this example, genus A has nine species, only three are distributed in India, while the rest are distributed in Southeast Asia and Africa. Let us assume that we know the relationships between these species (Fig. 11.6a). If only Indian samples were used to build the phylogeny of genus A, then we would get the tree as shown in Fig. 11.6b. This is an incomplete phylogeny as we have not included six other species from this genus. Furthermore, one might erroneously conclude that species A3 and A4 are closely related, which is not the case. Species A3 and A4 are related to two other species from Southeast Asia and species A7 is related to African species. Figure 11.6a also suggests that Indian members of the genus A are derived from Africa and Southeast Asia. This very interesting biogeographical conclusion cannot be drawn from Fig. 11.6b. Thus, incomplete taxon sampling has generated an uninformative tree that is not of much use and, worse, might give us the wrong information regarding species relationships if interpreted incorrectly! Many phylogenetic studies undertaken in India suffer from this problem because no thought has been put into robust sampling. Often, these studies are not publishable as the results are inconclusive. Papers by Bansal and Karanth (2010) and Kodandaramaiah et al. (2010) illustrate this issue in geckos and butterflies, respectively.

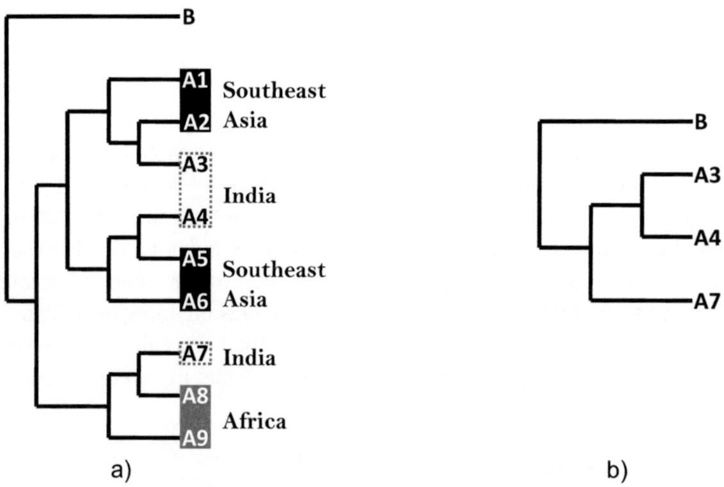

Fig. 11.6 Uninformative phylogenetic trees due to incomplete taxon sampling. The phylogenetic tree in (**a**) depicts the known relationships between species within a genus A and their respective distributions. (**b**) shows the tree built using only the Indian species, which is biogeographically uninformative

Incomplete taxon sampling can also sometimes generate erroneous phylogenetic relationships. For example, we might have sampled five out of the six species in a genus. Given that much of the species diversity in the genus has been covered, we would expect to generate a robust phylogeny. Nevertheless, the species left out might be important. In Fig. 11.7, including species A6 has significantly changed the topology of the tree. Figure 11.7c is the expected phylogeny in the absence of A6; instead, we get the phylogeny shown in Fig. 11.7a. Often bootstrap support values are informative in detecting such gaps in sampling. In this case, the bootstrap support for the node representing the common ancestor of A3, A4 and A5 in Fig. 11.7a might be very low (*see* arrow).

One (or more) of the species in my tree is not monophyletic. What does it mean?

Assuming that there is no contamination or mix-up of samples, polyphyly of a species could be due to multiple possible reasons. The first possibility is that the species of the group being studied have diverged from each other recently, due to which they exhibit incomplete lineage sorting at multiple loci (*see* Chap. 7). Secondly, some populations of a species might have hybridised with a closely related species, thereby rendering it polyphyletic at some loci. Thirdly, the polyphyly of a species might be due to erroneous taxonomy wherein unrelated lineages have been assigned to the same species (based on overall similarity), such as in the case of Hanuman langurs (see above). Such unrelated lineages that are morphologically indistinguishable from each other are called cryptic species.

Are gene trees always congruent with the species tree?

No, there are many possible reasons for incongruence between a gene tree and the species tree such as incomplete lineage sorting, random sorting of ancestral

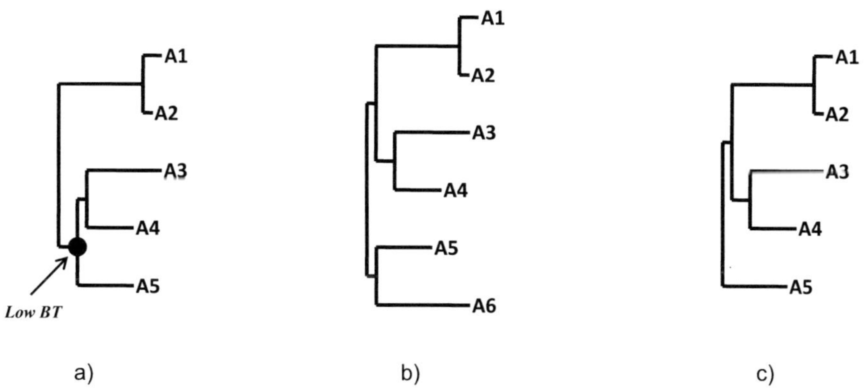

Fig. 11.7 Erroneous phylogenies due to incomplete taxon sampling. (**a**) depicts the erroneous tree retrieved due to the omission of species A6 (low bootstrap support shown as shaded circle on the node). (**b**) shows the phylogenetic relationships within genus A, with complete taxon sampling. The expected phylogeny in the absence of A6 is shown in (**c**)

alleles, horizontal gene transfer, hybridization (introgression) and positive selection (*see* Chap. 7, 'Gene tree vs. Species tree'). Therefore, multiple markers must always be used.

There are many gaps in my alignment. Is it okay? How do I treat gaps?

Gaps are introduced in the alignment due to indels (insertions or deletions) in one or more sequences. Indels are often very common in non-coding regions such as mitochondrial D-loop or nuclear pseudogene and sometimes also in coding regions such as in 12S rRNA and 16S rRNA. Thus, it is not unusual to have gaps in the alignment. Gaps can be used as a phylogenetically informative character if there is reason to believe so. Nevertheless, it is important that the gap is in the right place; otherwise, it might support erroneous relationships. Currently, only in the parsimony settings of PAUP (Swofford 2001) can one treat gaps as a fifth character; in the distance method, one can calculate total distance (without the use of models). Models of sequence evolution cannot deal with gaps, so likelihood and Bayesian inference cannot use gaps as an informative character (but see below). Gaps can also be treated as missing data in distance-based tree-building methods. In such cases, the programme ignores that character in all taxa (complete deletion) or only in comparisons where one of the members has a gap (pairwise deletion). For 12S and 16S rRNA dataset, it is advisable to treat gaps as missing data.

Gaps versus missing data: How to treat them?

A gap in a sequence alignment occurs when one or more sequences have an insertion or a deletion (indel). A gap of a single base in a sequence is coded as '-'. On the other hand, when sequences are not available for a region or a gene in an alignment, then it is considered missing data. Missing data are coded as '?' or 'N', meaning the base at that site for that sequence could be A/T/G/C (Fig. 11.8a, b). Likelihood-based methods (ML and BI) will consider all possible bases for the missing data while calculating the likelihood of that site for a given topology. This does not contribute any phylogenetic information. Parsimony, on the other hand, will impute a base that will optimise the parsimony score (tree length) of that site for a given topology. This can be problematic because parsimony might incorporate an incorrect base at that site which will in turn generate an erroneous score for a topology. For example, at a site with missing data, subsequent sequencing effort might reveal the site to be 'A', but the most parsimonious assignment at that site might be a 'G', which would depend on the tree topology and the distribution of bases at that site in other species in the alignment.

In distance-based methods, one can use two approaches, complete and pairwise deletion, to deal with missing data (these options might not be available in all programmes). In complete deletion, the column with the missing data is removed from the analysis, i.e. a particular site from all sequences in the alignment is removed. One can do this by asking the programme to delete a particular site in the alignment. Alternatively, one can use pairwise deletion, wherein the site is deleted only for the

```
     Gene1    Gene2
Sp1  ATCTACGG GTAGCCAT
Sp2  AGCTA??? GTAGGCAT
Sp3  ATCTA??? ??AGCGAT
Sp4  AGCTATGT GTAGCCTT
Sp5  AGCTCTGT GTAGCCAT
Sp6  AGCTAT?? GTTGGCAT
```

a)

```
     Gene1    Gene2
Sp1  ATCTACGG GTAGCCAT
Sp2  AGCTATGT GTAGGCAT
Sp3  ATCTACGG GTAGCCAT
Sp4  AGCTATGT ????????
Sp5  AGCTCTGT ????????
Sp6  AGCTATAT GTTGCCAT
```

b)

Fig. 11.8 Examples of missing data. (**a**) Data not available at the beginning or the end of a gene. (**b**) Data not available for a gene 2 from species 4 and 5

Fig. 11.9 Nucleotide deletion in species 2, no other variation in column 3

```
     12345678
Sp1  ATCTACGG
Sp2  AG-TATGT
Sp3  ATCTACGG
Sp4  AGCTATGT
Sp5  AGCTCTGT
Sp6  AGCTATAT
```

particular pairwise comparison where one of the sequences has missing data. Complete deletion option can also be used in likelihood and parsimony approaches.

What do we do if the gaps are real?

If the indel is seen in only one sequence, then use complete deletion option as this is a unique character change and is uninformative with respect to choosing between trees. In the example below, one can delete column 3 (Figs. 11.9 and 11.10).

However, complete deletion of the column with gap can be problematic if there is much variation at this site across sequences. This might result in loss of valuable information. In the example below, at column 3, species 2 has a gap, 3 and 4 have 'A', while 1, 5 and 6 have 'C' (Fig. 11.11). One possibility is to code this gap as missing data (?/N) instead of '-'.

However, if the gap is present in two or more species, it could be a synapomorphy. In parsimony, such gaps can be used as informative characters. In the

Fig. 11.10 Nucleotide insertion in species 3, no other variation in column 3

```
    12345678
Sp1 AT-TACGG
Sp2 AG-TATGT
Sp3 ATCTACGG
Sp4 AG-TATGT
Sp5 AG-TCTGT
Sp6 AG-TATAT
```

Fig. 11.11 Nucleotide deletion in species 2, variation in bases in column 3

```
    12345678
Sp1 ATCTACGG
Sp2 AG?TATGT
Sp3 ATATACGG
Sp4 AGATATGT
Sp5 AGCTCTGT
Sp6 AGCTATAT
```

programme PAUP, gaps can be treated as a fifth character, so we will have A, T, G, C and -; however, the datatype should be specified as 'symbols' and not 'DNA' (Fig. 11.12).

Alternatively, gaps can be coded as '0' for no gap and '1' for gap present. However, one should then also invoke the complete deletion option for the column with the gap (column 3 in Fig. 11.13). Here, the alignment has six characters, so we will have A, T, G, C, 0 and 1; also, the datatype should be specified as 'symbols' and not 'DNA'.

Nevertheless, if the column with gaps exhibits high nucleotide variation, then instead of complete deletion, one could code gaps as missing data (?/N) even though the gap is real (Fig. 11.14).

The 0/1 coding is also used when we want to combine molecular data with other kinds of data such as morphology and behaviour. Likelihood-based methods (ML and BI) cannot deal with gaps. This is because the models of sequence evolution used in these approaches are nucleotide substitution models that do not account for a change from a base to a gap or vice versa. Additionally, treating gaps as missing data does not contribute any phylogenetic information. Thus, if gaps are not informative, then it is best to code them as missing data in ML and BI analyses. Alternatively, go for complete deletion to save computational time. However, if one suspects that gaps are informative, then the binary coding (0/1) approach discussed can be implemented in MrBayes. A separate binary model needs to be assigned to

Fig. 11.12 Nucleotide deletion shared by species 2 and 3. No other variation in column 3

```
    12345678
Sp1 ATCTACGG
Sp2 AG-TATGT
Sp3 AT-TACGG
Sp4 AGCTATGT
Sp5 AGCTCTGT
Sp6 AGCTATAT
```

Fig. 11.13 Nucleotide deletion shared by species 2 and 3. No other variation in column 3

```
    123456789
Sp1 ATCTACGG0
Sp2 AG-TATGT1
Sp3 AT-TACGG1
Sp4 AGCTATGT0
Sp5 AGCTCTGT0
Sp6 AGCTATAT0
```

Fig. 11.14 Nucleotide deletion shared by species 2 and 3. Variation in bases in column 3

```
    123456789
Sp1 ATCTACGG0
Sp2 AG?TATGT1
Sp3 AT?TACGG1
Sp4 AGCTATGT0
Sp5 AGATCTGT0
Sp6 AGTTATAT0
```

this data partition. The programme BEAST can also treat ambiguous codes (Y, R, M, K, S, W) as informative characters, but to do this one needs to edit the XML file. Currently, most likelihood programmes cannot use gaps as informative characters. In the case of distance-based tree-building methods, as in parsimony, gaps can be coded as fifth character or as binary character (0/1). However, the distance measures that one can use are limited. None of the model-based distance measures can be used given that the data are no longer considered DNA sequence but just characters (symbols).

Some members of the ingroup are forming a separate clade and their phylogenetic position appears very unusual.

If one is working with mitochondrial sequences, then numts might have been erroneously amplified for the species in question (*see* Chap. 8; also Karanth 2008). Alternatively, the chosen marker might be a member of a gene family and one might have included paralogs instead of orthologs in the alignment (*see* Chap. 8). Reverse complement these sequences and realign. Look at your alignment very carefully. Check for sequences that appear very different from the rest.

How to deal with partial sequences?

Code them as missing data. The symbol '?' is universally used as missing data (Fig. 11.8).

There are many programmes for phylogenetic reconstruction—which one to use?

In the case of distance-based methods, as well as parsimony and likelihood methods, one can use PAUP (Swofford 2001) and MEGA (Tamura et al. 2021). The programmes RAxML (Stamatakis 2014) and IQ-TREE (Minh et al. 2020) are often used for ML analysis. For Bayesian analysis, MrBayes (Ronquist et al. 2012) and BEAST (Douglas et al. 2022) are widely used. The programme MEGA is very good for importing sequences from GenBank, sequence alignment and some data analyses, including preliminary phylogenetic analysis. For large datasets, other programmes might have to be used.

How do I describe a tree and interpret the results?

In many phylogenetic studies, the trees are not appropriately described, and often the results are misinterpreted. This is largely due to lack of understanding of what a phylogeny really represents. Hopefully, this book has clarified some of these issues and provided the conceptual framework for the correct interpretation of the phylogenetic analysis. While describing a tree, the first thing one should pay attention to is whether the tree is a cladogram or a phylogram or a phenogram or a time tree (dated phylogeny). Usually, a phylogram is reported as it is more informative (i.e. branch lengths correspond to number of changes). Time trees are useful when divergence dates are important to address the question, particularly in biogeographical analysis. Next, mention the major clades in the phylogeny and the members that make up these clades. If these clades are geographically restricted, then mention it as it might have interesting biogeographical implications. Similarly, report any other interesting finding such as unique traits shared by members of certain clades. Also, report any incongruence between the taxonomy of the study group and its phylogeny. Make sure you mention the support (bootstrap and posterior probability) for the node of interest.

References

Bansal R, Karanth KP (2010) Molecular phylogeny of *Hemidactylus* geckos (Squamata: Gekkonidae) of the Indian subcontinent reveals a unique Indian radiation and an Indian origin of Asian house geckos. Mol Phylogenet Evol 57:459–465

Douglas J, Jimenez-Silva CL, Bouckaert R (2022) StarBeast3: adaptive parallelized Bayesian inference under the multispecies coalescent. Syst Biol 71(4):901–916

Hillis DM (1995) Approaches for assessing phylogenetic accuracy. Syst Biol 44(1):3–16

Hillis DM, Moritz C, Mable BK (1996) Molecular systematics, 2nd edn. Sinauer Associate, Massachusetts

Karanth KP (2008) Primate Numts and reticulate evolutions of capped and golden leaf monkeys (primates: Colobinae). J Biosci 33(5):761–770

Karanth KP, Singh L, Collura RV, Stewart C-B (2008) Molecular phylogeny and biogeography of langurs and leaf monkeys of South Asia (primates: Colobinae). Mol Phylogenet Evol 46:683–694

Karanth KP, Singh L, Stewart C-B (2010) Mitochondrial and nuclear markers suggest Hanuman langur (primates: Colobinae) polyphyly: implications for their species status. Mol Phylogenet Evol 54:627–663

Kodandaramaiah U, Lees DC, Müller CJ, Elizabeth Torres K, Karanth P, Wahlberg N (2010) Phylogenetics and biogeography of a spectacular Old World radiation of butterflies: the subtribe Mycalesina (Lepidoptera: Nymphalidae: Satyrini). BMC Evol Biol 10:172

Minh BQ, Schmidt HA, Chernomor O, Schrempf D, Woodhams MD, von Haeseler A, Lanfear R (2020) IQ-TREE 2: new models and efficient methods for phylogenetic inference in the genomic era. Mol Biol Evol 37(5):1530–1534

Moore WS (1995) Inferring phylogenies from mtDNA variation: mitochondrial-gene trees versus nuclear-gene trees. Evolution 49(4):718–726

Ronquist F, Teslenko M, van der Mark P, Ayres DL, Darling A, Höhna S, Larget B, Liu L, Suchard MA, Huelsenbeck JP (2012) MrBayes 3.2: efficient Bayesian phylogenetic inference and model choice across a large model space. Syst Biol 61(3):539–542

Stamatakis A (2014) RAxML version 8: a tool for phylogenetic analysis and post-analysis of large phylogenies. Bioinformatics 9:1312–1313

Swofford D (2001) PAUP* – phylogenetic analysis using parsimony and other methods, version 4. Sinauer Associates, Sunderland

Tamura K, Stecher G, Kumar S (2021) MEGA11: molecular evolutionary genetics analysis version 11. Mol Biol Evol 38(7):3022–3027

MIX
Papier aus verantwortungsvollen Quellen
Paper from responsible sources
FSC® C105338

If you have any concerns about our products,
you can contact us on
ProductSafety@springernature.com

In case Publisher is established outside the EU,
the EU authorized representative is:
Springer Nature Customer Service Center GmbH
Europaplatz 3, 69115 Heidelberg, Germany

Printed by Libri Plureos GmbH
in Hamburg, Germany